ATTAINING FOOD SECURITY

ATTAINING FOOD SECURITY

Experiments of 10 Asian Nations

Alka Parikh

KALPAZ

ISBN: 978-93-5128-124-5

Price: ₹ 790

First Published, 2015

Published by

Kalpaz Publications
C-30, Satyawati Nagar,
Delhi-110052
Ph. : 9212142040
E-mail: kalpaz@hotmail.com

Cataloging in Publication Data—DK
Courtesy: D.K. Agencies (P) Ltd. <docinfo@dkagencies.com>

Parikh, Alka, author.
Attaining food security : experiments of 10 Asian nations / Alka Parikh.
pages cm
Includes bibliographical references.
ISBN 9789351281436

1. Food security—Asia—Case studies. 2. Food security—Government policy—Asia—Case studies. I. Title.

DDC 338.195 23

Contents

Acknowledgements

First of all, I would like to acknowledge Dr. Sachin Kumar Sharma of Centre for WTO Studies, Indian Institute of Foreign Trade, Delhi, for pointing out the need for a book on this topic. He said that there are no such reference books available that compare the food security policies of different nations. He explained how important it was for the Centre for WTO studies to have that information at hand to defend India's stand in WTO talks as well as to chart out the direction for Indian food security policies in the face of stiff international resistance. If he had not given that idea, this book would have never been thought of.

The compilation of data, without which the book would not have been possible, was done by the following ten B.Tech students of DAIICT: Sarthak Mohan Singh, Koosh Doctor, Sanchit Agarwal, Divy Thakkar, Rohan Sharma, Saharsh Yashlaha, Manas Roy, Shashank Jain, Nirav Devpura and Nikhil Pritmani. Summer internships are never taken too seriously by students but these students put in a lot of efforts and gathered enough material to put together this book. It was a new subject area for them, but that did not come in their way. They took a lot of interest and kept working till they completed gathering all the information needed to write the chapters. Thank you, guys, this book is dedicated to all of you.

Lastly, I want to thank my husband for tolerating my work schedule for five long months when I worked late nights, would not want to go out anywhere or pay much attention to household chores, I also want to thank my parents for understanding me when my regular phone calls became very irregular and my visits to Bombay decreased. Thank you to all you for always being there for me and giving such strong support.

PART I
INTRODUCTION

1

Strategies for Attaining Food Security: Lessons Learnt

Public Distribution System (PDS) for foodgrains is perhaps the largest welfare programme of India, if we do not count the petroleum subsidies (the subsidies are discontinued at present). It has a network of more than 5 lakh fair price shops (Press Information Bureau, 2011). According to the food security act passed in the parliament in 2013, the Government of India was planning to spend Rs. 1,20,000 crores to provide subsidised food to almost 80 crore beneficiaries (DEPD, 2013). This included not only the PDS network but also the school children for mid- day meals, *aanganwadi* children and expecting as well as lactating mothers. This is perhaps the largest experiment ever in the world to address the problem of hunger.

India needed to take some drastic steps towards achieving food security because the situation is grim. Forty two per cent of Indian children under the age of 5 are malnourished (World Bank Social Indicators). The IMR and MMR are high at 46 and 190 - they are comparable to the poor nations of the world; and much higher than the BRICS nations club that India belongs to. It is not even close to attaining the target set by the millennium development goals. In that sense, the food security act was a desirable legislation.

However, it is also known that PDS is fraught with numerous problems of inefficiencies and corruption. There is a dire need to think of alternative ways of supporting the population more effectively to save it from hunger. More cost effective and efficient systems need to be looked into. This book is an attempt to understand what other nations are doing to address the problem of food security and what lessons India can learn from them. For this, secondary data is compiled from different sources on different nations' policies towards attaining food security.

The study is of importance because there are no sources in our knowledge that give such comprehensive compilation of food policies across so many nations. This study is designed to act as a reference book that has all the relevant information compiled in one source.

Sample Nations

The nations are chosen mainly based on the geographical distribution. South Asia is known to be one of the most vulnerable regions for hunger. Almost all the nations of South Asia are selected for the study, including Nepal, Pakistan, Sri Lanka and Bangladesh. Since they all are considered vulnerable, it would be informative to know what is being done there to address the problem.

Asian tigers from South East Asia have fared much better than India in tackling their economic as well as hunger problems but there are some other nations who were ravaged by wars and could not do so well. Representatives of both such groups are included in the sample: Thailand and Indonesia represent the Asian tigers and Vietnam and Cambodia are the representatives of the nations that faced many political disturbances.

The nation most comparable to India in its population is China. It has attained much better levels of social development compared to India. India has to understand the strategies used by China, if it is looking for some practical answers. Mongolia is also in East Asia and is struggling to keep up its level of low hunger that it had attained during the Soviet influence. It shows how a nation helped and held by a communist nation handled the problem of hunger.

Thus 10 nations are chosen for the case study. The case studies start with listing out relevant indicators for each nation to understand the level of success achieved in attaining food security. Simple statistical tools like mean and standard deviation, regression and paired t test are used to interpret the data.

The strategies for attaining food security vary a lot and are influenced considerably by the region's history. A brief history is given in section III for each nation to contextualise the current policy approaches. The current policies for attaining food security are discussed in details in the section following history. If possible the budget for such policies is also given. Critical evaluations of such policies are also given whenever available.

All the above data is compiled from the secondary sources available from the web.

II. Approaches of Different Nations

It is interesting that out of the 10 nations that we studied, except India, few have even tried to reach their poor with subsidised food and at present, no one has any such programme for its poor except Indonesia. Bangladesh and China gave up the programme after realising that the leakages are too many and hence the cost of reaching the poor was coming out to be very high. Midday meals schemes for the school children are also working only in five nations out of 10 nations. And yet, six nations had much better health and nutrition indicators compared to India, and the remaining four had rates that were comparable to India.

This section summarises the approaches used by different nations in addressing the problem of malnutrition.

Sri Lanka

Sri Lanka has achieved a lot on the social development front in spite of being a developing nation. It is a poor nation but its human development index is comparable to middle income countries. This is because in the initial periods of its development after independence (1948–1977), Sri Lanka invested almost 30-40 per cent of its public expenditure on social sectors (UNRISD, 2007).

Like many nations, Sri Lanka also abolished its subsidised food distribution programme and instead introduced the innovative food stamp programme. But it has been found that after the introduction of the programme, the bottom 20 per cent actually decreased their calorie consumption. This is because the food stamps did not reach the most vulnerable. Many non-eligible households got included in the programme instead. Also, the nominal value of food stamp was kept the same for a long time while the prices were increasing. Thus, the real value of the food stamp almost reduced by half. The impact that it could have on food security also became marginal (Edirisinghe, 1987; Josling, 2011).

The situation thus must have deteriorated a little after the welfare expenditures were cut down. We could not find a continuous series to establish this negative impact. However, the fact remains that malnutrition rates have increased a lot in Sri Lanka now due to the war for many years. In the war ravaged north, 50 per cent of the children are found to be malnourished. (Amaranath and Fernando, 2010). Still, inspite of all such changes, the social development indicators have remained far better than India even today. The positive effect of heavy investment made in social sectors lingers on.

At present, Sri Lanka is concentrating more on maternal and child health through its various food distribution programmes since these are found to be the most effective in addressing the malnutrition problem in the country. The provision of micro nutrients and other support is also made through these programmes. A programme similar to food for work programme is used to increase rural incomes and gives part payment in terms of food stamps.

Indonesia

Indonesia is a lower middle income country. Indonesia's per capita income was almost 2.5 times the per capita income of India. Poverty mitigation efforts in Indonesia have shown meaningful progress, which has been in accordance with the MDGs as was demonstrated by the reduced proportion of people living under the national poverty line, i.e. from 15.10 per cent (1990) to 11.66 per cent (2012). Between 1989 and 2010 the prevalence of under-five children with low weight also went down from 31 per cent to 17.9 per cent (UNDP, 2011).

To ensure food security for its urban consumers in 1950s upto 1970s, Indonesia experimented with keeping the price of rice low for rationing. Indonesia was not self-sufficient in rice hence this practice proved to be a major drain on the foreign exchange of the country.

The rice production in the main area was declining due to bad maintenance of the irrigation system. It was advocated that the country would do well to import fertilisers and maintain irrigation systems rather than keep importing expensive rice. But it was found very difficult politically to increase rice prices.

Later, it was decided to support farmers through irrigation, fertiliser subsidy, subsidised credit and floor price but the failure of monsoon upset all plans. So, farmers' encouragement and rice imports were carried on simultaneously from then onwards. But soon the production started rising. By mid 1980s, the surplus stocks spilled over the warehouses and it was impossible to keep the floor prices alive. As the share of rice declined in the national income, the benefits from rice stabilisation decreased but the cost to the government was increasing. So it was decided that imports would be used as balancing factor rather than buffer stocks to stabilise rice supplies and prices. Government monopoly over buying and importing rice was terminated. The new flexibility in rice trade helped in reducing costs (BAPPENAS, 2002).

Three important current programmes aimed at attaining food security are as follows: (a) In the face of Asian financial crisis, Raskin was introduced to distribute rice to the vulnerable households at subsidised rates. Almost $ 134 billion were spent every year on this programme. The problems of food reaching the non-eligible households exist, but still, many households have been helped under this programme. (b) In 2005, the fuel subsidies were reduced and the ones to be affected by such cuts were given cash to meet the increased fuel bills. More than one third of Indonesians received this cash and it is reported that the transfers were more than enough to cover the increased fuel costs and during this period, the consumption of other things also did not decrease. The administrative costs of this cash transfer schemes were very low. It is an example that needs to be studied further so that it can be extended to other nations. (c) The conditional cash transfers are made to the woman of the family if the children are going to schools, if the delivery is made in hospitals, etc. This programme has also reached most of the chronically poor families (World Bank, 2012). In conclusion, in spite of its unsuccessful attempts with rice subsidisation, Indonesia has finally attained its goals with the help of increase in incomes and by effectively implementing a few direct welfare programmes.

China

China's growth story is non-comparable. It has grown much faster than almost any other country, especially given the huge population that it had to carry along. Yet, its inflation has been kept under control, even during the years of depression. Its malnutrition and health indicators are almost comparable to any developed nation (CIA World Fact Book, 2011).

China has been able to attain highest yield per hectare in many crops that it specialises in, showing how well its agriculture sector has grown. But it was not always that way. In 1956, about 100 million farmers were suffering from shortage of food. Even in 1978, China was not self-sufficient in food. In 1980s, slowly the farmers were given more rights over land. The government invested a lot in agriculture research and its extension. The state monopoly in buying and selling grains was removed and farmers were allowed to sell their products in the local market (Zhou and Gandhi, 2007). The food for work programme was used by the local governments to build rural infrastructure. Even if the farmers did not get paid for these programs, they participated in it, because it led to considerable rise in agriculture productivity as well as prices received for land and the produce.

To attain food security, China first achieved self-sufficiency in food. Increased domestic production increased the per capita food availability. This ensured food for its rural population because almost 85 per cent had land (He Qinglian, 2011). The urban population was supported by the local governments by providing subsidised food as and when required, although of late the government and open market prices are often the same. The rationing system of food has been substituted with cash transfer programmes and heavy administrative interventions for maintaining the grain prices at a reasonable level.

Decreasing dependence on agriculture by creating jobs outside the sector was another strategy that worked well. As surplus labour shifted out of agriculture, incomes for everyone increased. Diversification of production increased farmers' incomes and food security. Except for school feeding programmes, China now does not indulge into any food distribution programmes. Its aid to the poor continues through cash transfers. However, for the cash transfer programmes to work, control over prices is very essential. China has liberalised but still, the government exerts strong control on the economy (Ling and Zhongyi, 2004).

Vietnam

Vietnam has been growing at much faster rate than India. In 1994, Vietnam's per capita income was less than India's. In spite of impressive growth rates of India, Vietnam still surpassed India by 2013. In all social indicators also, Vietnam is far ahead of India (World Bank, 2013). One reason can be the communist rule in Vietnam that must have provided for basic amenities for all. Starting from such strong base of equality in the society, Vietnam must have found it easier to grow fast. Thus investing in basic amenities at the cost of giving good quality goods to the population, as done by the communist nations, might be paying off well in the earlier stages of development. The priorities in expenditure thus can be said to be better designed for achieving better quality social development.

When Vietnam started liberalising its economy under *doimoi* from 1986, the increase in incomes has been phenomenal. For food security, the nation has concentrated on giving incentives to farmers by leasing out lands to them and changing property rights in favour of the farmers. In addition, cash support, and price support were provided. Infrastructural facilities, mainly in the form of irrigation, storage, credit and agriculture extension are also given (Nguyen, 2006). Basically, the entire emphasis is on improving agriculture productivity. Since the extent of landlessness is very low, increased agriculture productivity is automatically supposed to lead to better nutrition for the households. We feel that since Vietnam started off from the foundations of a communist regime, inequalities in incomes were much lesser and the general demographic development levels were similar across most households. Also, enough land was available to distribute to all the people who desired to cultivate. Such ownership entitlement seems to have helped a lot. Therefore it was easier for the gains of growth to reach everyone, thus leading to large scale improvement in malnutrition levels.

Thailand

Thailand is an upper middle income nation in South East Asia. It was ranked second in the best quality of life in ASEAN. Thailand has developed from a low income nation to a middle income nation in last 50 years. The deadline for achieving millennium development goals is 2015. Thailand is already on track to meet almost all indicators including poverty eradication, high education rates, reducing child and maternal mortality and providing clean water and sanitation facilities. Naturally, in all indicators of social development, Thailand is far ahead of India.

Thailand has managed to weather political uncertainties, disastrous tsunami and the Asian crisis, to maintain its high status in social development. Its infant mortality rate is just eleven, poverty head count is 13.2 and malnutrition in proportion of children below the age of Five (weight/age) is Seven per cent (world bank social indicators).

Since the absolute figures and macro realities are not so worrisome, for many years, Thailand has concentrated on the detection of the problems of malnutrition and addressing them right at the source. Since 1960, Thailand conducts national nutrition surveys (now known as food and nutrition surveillance system-FNSS).. FNSS attempts to identify any shortfalls and danger signal so that immediate action can be taken. Food coupons are given for the affected children from which the mother could buy nutritious food every day (NESDB, 1992). School lunch programme also tries to provide supplementary nutritious food. No other direct feeding programme or subsidised food programme for the population at large exists. For reducing poverty, some income generation programmes for developing community assets are implemented together with provision of basic services and strengthening of rural services infrastructure. The universal pension scheme takes care of the old and invalid who cannot work. Universal health coverage provides free health care for all diseases and covers 74 per cent of the population. This has prevented families from falling into poverty due to reducing instances of catastrophic health expenditures (Sakunphanit and Suwanrada, 2011).

In short, Thailand has addressed its malnutrition problem by providing food to the children, by giving pensions to the old and by providing universal health care. Income generation programmes for building community assets are also used. When the general well-being is attained by high growth for most of the people, such focused efforts seem like a good strategy.

Nepal

Nepal is a poor nation, just like India. Its social development indicators are also similar to that of India. It almost feels as if it shares the same fate as India in every way. Most of the population is heavily dependent on agriculture. The farm population has also been growing at a higher rate than the rest of the nation. This, together with limited scope for expanding agriculture and extremely unequal land distribution, has made the region poorer over the years.

Nepal has a public distribution programme for providing subsidised food but it works only in a limited area. It is very much appreciated where it is implemented but due to its limited reach, its impact on overall food security situation is not much. Nepal does not have funds of its own to implement any significantly large PDS programmes. World Food Programme (WFP) has provided some aid under its country programme but it is operative only in some parts (WFP website). People need help here but the government does not have money to provide any support.

Bangladesh

It has been growing at the rate of 5.89 per cent p.a. in last decade (World Bank social indicators). The gains in incomes came from liberalisation policies that affected agriculture also positively (mainly in terms of imports of rice and inputs), among other things. Such gains have been reported to have translated into more income for labour (World Bank, 2013), leading to fall in poverty and malnutrition levels. However, the change is not enough and Bangladesh is far from achieving the millennium development goals.

The strategy for reducing poverty and malnourishment figures has been accelerated growth. Agriculture growth happened in the 1980s and 1990s by adopting HYV technology and also because farmers started investing in private tube well. This irrigation infrastructure together with HYV technology led to almost tripling of the production in last 30 years (Hagblade, 1994).

In addition, import and export markets were liberalised. Government system suffered from many inefficiencies, so the imported rice and chemical inputs were more expensive when it was imported by the government. Due to lesser price of imported rice and the lesser input prices, the real price of rice actually came down after liberalisation. Thus both availability and affordability increased (Hagblade, 1994).

It should be noted that the government did not have much role to play in bringing down either the poverty or the malnutrition figures. Government runs some small programmes but not at a scale that can make much difference. All the programmes that we came across covered not more than Three-five per cent of the population. The international aid has also been declining. Thus the fall in malnutrition figures till now seems to have been achieved mainly because of growth in the economy. However, more aggressive approach would be needed if the figures were to fall faster and to meet the world standards of nourishment.

Cambodia

Cambodia is one of the poorest countries in south East Asia. It did grow at the rate of 9.5 per cent p.a. from 1999 to 2008, but the population grew at 2.3 per cent p.a. in the same period. Also the people active in agriculture grew by three per cent. As a result, when the per capita grew at 6.3 per cent annually, the value added per agriculture worker grew by 1.9 per cent only. Due to this, the income inequality worsened. Cambodia has the fastest growing income inequality in the South-East Asian region today (Tickner, 1996).

Due to this, the performance of Cambodia has been quite unsatisfactory on social development front. The indicators are improving, but a lot still remains to be done.

Interestingly enough, Cambodia also has taken a different route to achieve better indicators. For attaining food security, it emphasises increase in food production and agriculture growth but does not talk about price support for farmers. The increase is to be achieved through irrigation and technology - that too, technology that is low cost, like SRI. It plans to increase access to land through distributing de-mined land to the rural poor.

Access to food is to be achieved by creating employment for the poor mainly by providing training, help in developing micro enterprise and spreading information regarding markets. Direct employment programmes by the government are also only a few. There is no mention of any subsidised food, except in emergency situations. For children, direct feeding programme is not on plans, mainly the emphasis is on spread of awareness rather than provision of food. Fortified food and micro nutrient supplements are mentioned but on a small scale.

In short, no large scale programmes are planned for improving food security. Yet, till now, due to large number of deaths in the disturbed period, Cambodia has relatively less density of population - 204 persons per square km (U S census Bureau, 2007). The country can afford to concentrate on proper distribution and utilisation of resources for achieving increase in incomes and food security. But as population is increasing rapidly, its problems can multiply, if the remedial measures are not taken quickly.

Pakistan

Pakistan trails behind India in all the economic and social indicators. The malnutrition figures somehow are shown to be lower but they seem to be contradicting with the overall picture presented. In general it is safe to conclude that Pakistan's social development seems to be at a lower stage than India (World Bank Indicators).

In spite of such low indicators for social development, Pakistan government's policies do not seem to be directed at food security. The wheat purchase programme, that is supposed to be giving remunerative price to the farmers and protection to the consumers, has never been implemented properly. It seems to be failing on both counts, benefiting none.

Zakat programme, wherein the government imposed a 2.5 per cent tax levy (a wealth tax) on specified financial assets of individuals such as bank deposits worth more than a certain amount. The programme operates under the principle that deserving needy (Mustahqieen) Muslims are to be given assistance. This particularly includes widows, orphans, the disabled and the unemployed (but not beggars). It was an interesting experiment, conducted in the true Islamic spirit. The funds were used to help the needy in a variety of ways, although food provision was not the main aim. However, after the programme became voluntary, the funds are drying up (Khan and Qutub, 2010). Since it was not focused on just one target group, there was little macro impact generated.

Interventions related to nutrition have been implemented such as vitamin A supplementation, wheat flour fortification and salt iodisation, but these have largely been stand-alone programmes that have been implemented in isolation rather than as a complementary nutrition package. Support for mother and child health has been partially addressed through the Lady Health Workers scheme introduced by the government to promote good health and provide basic services to women at village level. The overall humanitarian nutrition response has focused primarily on short-term emergency treatment programmes for acute malnutrition with little focus on addressing the underlying causal factors and on establishing long-term, sustainable and preventive programmes through the provincial and national governments.

National Zero Hunger Action Plan was an ambitious project to reach 61 million hunger affected people but the programme has not started even after seven months of its inauguration. Special Program for food security I and II, an ambitious programme that aimed to enhance agriculture productivity, is abandoned because the central government refuses to provide funding for it, since it is a provincial subject (Inpaper, 2012). In short, Pakistan does not have any proper programmes for food security. Per capita food availability fluctuates widely and the population is always in the danger of becoming food insecure. The improvement in malnutrition indicators is rather slow. The disturbances in the country can worsen the food security situation.

Mongolia

Since 1924, Mongolia was under the Russian influence. It was a Soviet satellite state. After 1990, when USSR disintegrated, Mongolia got its freedom. It is a lower middle income country.

Under the Soviet rule, Mongolia received a lot of support, food aid and capital aid from USSR. Its social indicators were good. Literacy was high, health indicators were satisfactory and the hunger levels were very low. After independence, Mongolia had to fend for itself. That has not been easy. Hardly 1 per cent of Mongolia's land is arable. Without Russian support, most state owned enterprises could not survive. The farmers did not have the resources to improve their agriculture and hence agriculture production has been slowing down. Import of food from China has been becoming more and more expensive (FAO Document Depository). Inflation and malnutrition have been rising. Efforts are on to improve food production and attain self- sufficiency in food. However, their journey has just begun so there are many road blocks. The income from mining can keep the economy going by not only creating jobs but also by earning foreign exchange to buy food from other nations. But at present, the exports are much lower than imports and Mongolia's social indicators also can go down, if the situation continues to remain the same.

III. Analysis of Major Strategies used by Nations to Tackle the Problem of Food Security

Economic Growth

In case of Thailand and Indonesia, economic growth and increase in general productivity led to overall fall in poverty and malnutrition. It is seen that achievements attained by high growth rates are also sustainable over a period of time. However, growth does not always lead to decrease in food insecurity. It is not a solution that would always work. India has been growing very fast and yet the food insecurity did not decrease because income growth was experienced only by a limited group. The rest of India did not get to share the benefits of the growth. Similar is the case of Cambodia. In such situations, the government should take steps to redistribute at least some part of the gains from the growth by increasing spending on programmes for the vulnerable.

Agriculture Growth

Increasing the production and productivity of the farmers is the strategy that many nations are using to address the issue of food security. Bangladesh, China and Vietnam tried to remove the restrictions and regulations for agriculture and liberalised the institutional set up for increasing production. It actually worked for them. The productivity increased many folds.

Nations like Vietnam and China that concentrated on agriculture development as the solution for food security found that as production increased, the food insecurity reduced considerably. However, it should be understood that when the landlessness is much higher as in the case of India and some other South Asian nations, mere agriculture growth would not work. In China, only 15.5 per cent are landless, in Thailand 14 per cent and in Vietnam, the number is 12.3 per cent. In these nations, governments succeeded in increasing the agriculture productivity through technology dissemination or liberalisation. Increase in productivity automatically led to more production of foodgrains and thus lesser food insecurity for the rural households-because most households in rural areas were farmers with land. In India, where almost 43 per cent of the rural farmers are landless, increase in agriculture productivity leads to better conditions for farmers with land. But it does not guarantee higher food security for all agriculture labourer households.

That is why increasing agriculture production and attaining self-sufficiency in food did not help India in attaining food security for all. As shown in table 1, India has much less land available per person compared to most of the South East and Central Asian nations. It thus does not have the option of giving out land to the landless and teach them farming technology to grow food for themselves. If agriculture growth is so high that it results in reduction in real prices of the food grains, the gains from such growth reaches all because now food is more within reach for many more families. This happened to an extent in Vietnam and Bangladesh.

Table 1: Population, Land Area and Density of Population of the Sample Nations

Country	Population	Land Area (sq mi)	Density per sq. mi
Bangladesh	147,365,352	130,168	1132
India	1,095,351,995	1,147,949	954
Sri Lanka	20,222,240	24,996	809
Vietnam	84,402,966	125,622	672
Pakistan	165,803,560	300,664	551
Nepal	28,287,147	52,819	536
China	1,313,973,713	3,600,927	365
Indonesia	245,452,739	699,548	351
Thailand	64,631,595	197,595	327
Cambodia	13,881,427	68,154	204
Burma	47,382,633	253,954	187
Oman	3,102,229	82,031	38
Mongolia	2,832,224	604,247	5

*(**Source:** U.S. Census Bureau, International Data Base)*

As food availability translated into food security for the South East Asian nations (except Cambodia), they could concentrate more on the quality of food ingested, like supply of micro nutrients. The figures of malnourishment are comparable for Nepal, Bangladesh, Pakistan, India and Myanmar. Except India, no other nation has undertaken large scale welfare programmes for attaining food security. The above table shows that given the large population, India has to try harder than any of its neighbor to attain even the levels that those nations stand at. The proportion of the problem explains the lack of difference in indicators in spite of having put in more efforts.

Liberalisation

When China and Vietnam allowed farmers to have some ownership rights over land and total rights over their produce, the productivity grew rapidly. Attempts of collectivisation of agriculture have never been very successful.

Government's monopoly rights over purchase and sell of grains have also been proved to be counter productive in these nations. Liberalisation accelerated their growth process. Opening up of markets and loosening controls over imports and exports helped the Bangladesh economy grow faster.

Indonesia also could stabilise its economy better with liberalisation of imports. Thus liberalisation seems to lead to better growth in at least some of our examples. However we would like to point out that the communist nations with their strict centralisation of power suffered when it came to increases in productivity (it could not be achieved) but gained when it came to achievements in social sectors. Control over all the national resources generated contradictory trends on economic and human development fronts.

Liberalisation might not be a preferred strategy always. Having some controls over markets has been deemed necessary at times. India and China still have strict controls over import and export of foodgrains because they have to ensure that the domestic demand is met fully before exports are made and that the farmers always get remunerative prices for their crops even if at times the crops are produced at prices higher than the international markets (hence restrict the imports).We feel that such control over markets might have helped them in remaining self-sufficient in food.

Price Control

Countries can either provide subsidised food, because the poor cannot afford food grains at the market price or they can attempt the keep the prices low so that no subsidies need to be provided. Experiences of different nations have shown that it has not been very easy to maintain prices at low levels or at steady levels. Imposition of certain prices has often led to shortages and black marketing, as the experience in India shows. Implementing low prices for the whole country is also very difficult as has been demonstrated by the experiences of Vietnam.

Especially, after liberalisation, as the agriculture markets have connected to the world markets, it has become more and more difficult for nations to maintain steady prices. Indonesian experience shows that when the country is not self- sufficient in food, and has to depend on the world market for supplies, maintaining steady prices becomes very expensive option. China also had to use force to maintain prices. Thailand went through the disastrous experiment of rice pledging scheme, trying to provide higher prices to the farmers by selling rice at a higher price in the international market – the government fell. Thus while economically it is an attractive option to maintain steady prices, countries have found it difficult to do so. That is why subsidies are needed to be provided to the poor : to enable them to purchase food.

Investing in Social Sectors

Investing in social sectors by devoting a considerable part of national income to them has been tried out by Sri Lanka and many communist nations. The results are noteworthy. Their indicators have been always considerably better compared to nations with comparable incomes. Ensuring basic amenities to all the citizens has worked wonders for Vietnam, China, Mongolia and Sri Lanka. China and Vietnam have also been able to leverage on these foundations very well. Sri Lanka and Mongolia are still struggling to stabilise their economies but the strong foundation of social sectors is expected to help them in going ahead. Prioritising expenditures in favour of social sectors can yield high returns in the long run. Of late, while defining food security, a holistic approach of well-being is used wherein balanced diet, clean water, sanitation and access of health care services are also included in addition to adequate quantity of food. When we look at food security in such a holistic sense, investing in social sectors becomes extremely important.

Self-Sufficiency in Food

It is shown by all the nations under study that providing subsidised food has not been possible unless the nation is self-sufficient in food or it has earned enough foreign exchange to import food. India could start its public distribution of subsidised foodgrains programme only after it attained food self-sufficiency. China also manages its price control in some subsidised food programmes well because it is self-sufficient in food.

On the other hand, Mongolia is struggling to attain food self-sufficiency because it cannot easily afford to import food. When it was under Russia, that problem did not exist because Russia was self-sufficient in food. Even when its levels of malnourishment are increasing, Mongolia with its meagre resources and inadequate food production does not have the choice of distributing food at a lower rate. The same is the case with Pakistan, Nepal and Bangladesh. They are not yet fully self-sufficient in food and hence cannot provide foodgrains at subsidised rates, even when their levels of malnourishment are very high. When nations have already accumulated enough wealth, like Oman or to some extent Indonesia, they can decide to import food as and when needed and concentrate on earning foreign exchange from their other remunerative activities.

IV. Conclusion

This paper identifies the effective strategies used by different nations in attaining food security: (a) Economic growth and agriculture growth have been the most dominant strategies tried out and they have worked well in many cases. However, whenever inequalities are high, the growth strategies have fallen short. (b) Removal of excessive controls on agriculture has helped some former communist nations in accelerating agriculture growth and also in attaining food security. (c) At the same time, most former communist nations and some other nations like Sri Lanka who invested heavily in social sectors like medicine and education have attained good indicators of well- being.

Their populations are much better off in nourishment levels. This strategy is effective in long term. (d) Price controls have not proved to be successful for most nations. (e) Achieving self-sufficiency in food has been a pre-condition for many nations for attaining food security but some nations that have earned large incomes from other sources have been able to attain food security by importing food grains.

In short, investing in social sectors and agriculture growth seem to be some of the best practices for attaining food security. Once the population is empowered with skill and good health, it can weather the adversities, even when the government support is withdrawn. Similarly, knowledge of technology would bring in benefits for a long time. These two factors can make a considerable difference in general well-being of a country, including food security. In addition, concentrating on feeding infants and young children as well as expecting and lactating mothers can also lead to reduction in malnourishment levels. Here surveillance system like that of Thailand can help in addressing the problem more effectively.

V. Lessons for India

Both economic growth rates and agricultural growth rates have been satisfactory for India. While other nations were struggling even in 1990s and 2000s, India achieved self- sufficiency in food by late 1970s. Economic growth reached new heights after liberalisation in 1990s and India started being considered as an emerging economy - a force to reckon with. Yet the figures for malnutrition and infant/child/maternal mortality rates remain high. This has happened because the growth has not reached all the citizens and regions of India. There are many regions that are still very underdeveloped, there are many workers who are still very exploited and underpaid.

Large size of population and lack of resource ownership or skills add to the problem. It is not that efforts have not been made to redistribute the gains from growth, but the policy implementation has been ineffective.

We need to concentrate more on underdeveloped regions to address the agriculture growth issues and on training the workers for improving their standards of living.

India has never concentrated its efforts on social sectors. This neglect has led to poor health and education indicators. That in turn has resulted into lack of employment, poor pay and widespread malnutrition. Case studies of all nations from East Asia and some from South East Asia as well as Sri Lanka point to this gap as the most important cause for malnutrition.

Should India continue with its highly inefficient and costly PDS to make subsidised food reach its poor? It is true that the system is very poorly implemented but at present, it is the only system in place that reaches so many Indians.

The alternatives of cash transfer, food stamps etc., have been considered but not found to be adequately satisfactory solutions. While experiments with other alternatives should continue, we must continue to try to improve PDS till we find a better solution. It does not make sense to abandon a system already in place, with nothing to substitute it.

As far as policy innovations are concerned, India seems to have tried out many policies for addressing the problem of hunger. Among all the nations studied, it seems to be a nation with highest number of schemes for food distribution and numerous policy innovations. The problem does not seem to be so much the lack of policies or of research studies as that of the implementation. Until India gets its governance and management right, it would be difficult for it to manage the problem of hunger.

References

1. BAPPENAS (2002) : "Food Security in an Era of Decentralisation: Historical Lessons and Policy Implications for Indonesia", February 2002 (http: /pdf.usaid.gov/pdf_docs/PNADE883.pdf)
2. CARD (2008), "Strategic Framework for Food Security and Nutrition in Cambodia 2008-2012", with Technical Working Group for Food Security and Nutrition, May 2008 http://www.gafspfund.org/sites/gafspfund.org/files/Documents/Cambodia_4_of_16_ STRATEGY_ Food_Security_and_Nutrition_Strategy_0.pdf
3. CIA World Factbook (2011): http://www.indexmundi.com/g/g.aspx?v = 30 &c = ch&c = in&l = en
4. DFPD, "National Food Security Act, 2013" http://dfpd.nic.in/?q= node 955
5. Edirisinghe Neville (1987), "The food stamp scheme in Sri Lanka: Costs, benefits and options for modification", Research Report 58, IFPRI, March 1987 http://www.ifpri.org/sites/default/files/pubs/pubs/abstract/58/rr58.pdf
6. Expat Focus: Oman Social Security and Welfare (http://www.expatfocus.com/expatriate-oman-social-security)

7. FAO: Corporate Document Repository (http://www.fao.org/docrep/005/y2722e/y2722e0y.htm)

8. Haggblade Steven (1994), "Evolving food grain market and food policies in Bangladesh" USAID, May 1994 (http://pdf.usaid.gov/pdf_docs/PNABS137.pdf)

9. He Qinglian (2011),"China's 100 million Landless Peasants", The Epoch Times, February 4, 2011 http://www.theepochtimes.com/n2/opinion/chinas-100-million-landless-peasants-50605.html

10. InPaper Magazine (2012), "No progress on Zero Hunger Project", December 9, 2012 http://www.dawn.com/news/770158/no-progress-on-zero-hunger-project

11. Josling Tim (2011), "Global Food Stamps: An Idea Worth Considering?", ISTSD Issue Paper No 36, August 2011, http://www10.iadb.org/intal/intalcdi/PE/2011/08841.pdf

12. Khan Shanza N. and Sara Qutub (2010), "The Benazir Income Support Programme and the Zakat Programme - A Political Economy Analysis of Gender", Overseas Development Institute, November 2010 (http://www.odi.org/sites/odi.org.uk/files/odi-assets/publications-opinion-files/7247.pdf)

13. NESDB (1992), "Strengthening Food and Nutrition Surveillance for Nutrition-Oriented Development Decisions" 1991 Progress Report, UNICEF, Thailand 1992.

14. Nguyen Viet, Tim McGrath and White Pamela (2006), "Agricultural Land Distribution in Vietnam: Emerging Issues and Policy Implications".(http://mpra.ub.uni-muenchen.de/25587/1/MPRA_paper_25587.pdf)

15. Press Information Bureau, Government of India, 2011 http://pib.nic.in/newsite/erelease.aspx?relid=74180.

16. Sakunphanit and Suwanrada, (2011) "Thailand 500 Baht Universal Pension Scheme" in Sharing Innovative Experience: Successful Social Protection Floor Experiences by ILO, 2011 http://webcache.googleusercontent.com/search?q = cache:t 01Vw Kahlb8J: www.social-protection.org/gimi/gess/Ressource Download. action percent 3 Fressource.ressourceIdpercent 3D24378 + & cd = 2 & hl = en&ct = clnk & gl = in

17. SujeewaAmaranath and Kapila Fernando (2010), "Sri Lanka: Widespread malnutrition among women and children", World Socialist Website, June 10, 2010 http://www.wsws.org/en/articles/2010/06/slan-j10.html)

18. Tickner Vincent (1996), "Food Security in Cambodia: A Preliminary Assessment", Discussion Paper 80, UN Research Institute for Social Development, October 1996http://www.unrisd.org/80256B3 C005BCCF9/(httpAuxPages)/64B8DCC5 49D852498 0256 B67005B6C36/ $file/dp80.pdf

19. U.S. Census Bureau, International Data Base (2007), "Population Density PerSqaure Mile of Countries", quoted by Information Please Database, Pearson Education, Inchttp://www.infoplease.com/ipa/ A0934666.html

20. UNDP (2011), "Report on achievement of millennium development goals Indonesia", 2011

    http://www.undp.or.id/pubs/docs/Report percent 2 0onpercent20the percent20Achievement percent20ofpercent 20thepercent 20MDGspercent20inpercent20 Indonesiapercent 202011.pdf

21. UNDP: Millenium Development Goals, "Eradicate Extreme Poverty and Hunger: Where do we stand?" (http://www.undp.org/content/ undp/en/home/mdgoverview/mdg_goals/mdg1)

22. UNRISD: Policy and Development (2007)," Development Strategies, Welfare Regime and Poverty Reduction in Sri Lanka", Country Overview Paper No. 7, Institute of Policy Studies of Sri Lanka. November 2007http://www.unrisd.org/80256B3C005BCCF9/search/ 684D9493D25F9EC6C1257AE8005CAC3B?OpenDocument

23. WFP (World Food Programme),"Food Security Atlas for Nepal"http:/ /www.foodsecurityatlas.org/npl/country

24. WFP: Current Operations: Myanmar(http://one.wfp.org/operations/ current_operations/project_docs/200299.pdf

25. Winichagoon P and Schuftan C. (1991), "Thailand: Food and Nutrition Surveillance System: :An Inquiry into Output Data Utilisation for Decision-Making", UNSSCN, 9-15 December 1991

26. World Bank (2012), "Public Expenditure Review Summary" Working Paper 67324, February 1, 2012

27. World Bank (2013) "The World Bank Report", 2013 (http://data.worldbank.org/country/vietnam)

28. World Bank Social Indicators http://data.worldbank.org/indicator/NY.GDP.MKTP.KD.ZG/countries/BD?display=default

29. World Bank social Indicators, http://data.worldbank.org/indicator/SH.STA.STNT.ZS/countries

30. World bank: News, "Bangladesh Poverty assessment: A decade of progress 2000-2010", June 20, 2013 http://www.worldbank.org/en/news/feature/2013/06/20/bangladesh poverty-assessment-a-decade-of-progress-in-reducing-poverty-2000 2010

31. Zhou Zhang-Yue and Vasant P. Gandhi (2000), "Public Food Distribution System and Food Security of the Poor: The Indian and Chinese Experience Compared", Paper presented at the 44th Annual Conference of the Australian Agricultural and Resource Economics Society, the University of Sydney, Australia, 23-25 January 2000. (http://ageconsearch.umn.edu/bitstream/123746/2/Zhangpercent20Yue.pdf)

32. Zhu Ling and Jiang Zhongyi (2004), "The Food for Work Policy and Expansion of Rural Employment in Poor Areas in Western China", Employment Intensive Investment Branch, ILO, Geneva (http://www.ilo.org/public/english/employment/recon/eiip/download/setp/setp16.pdf)

Part-II
NATIONS WITH BETTER FOOD SECURITY

2

Sri Lanka

Compilation: Sarthak Mohan Singh

Introduction

Sri Lanka has been always praised for its achievements in human development despite being a developing country. In spite of having a low per capita income and low levels of economic development, its social indicators are similar to those of middle income nations. And its achievements in social development were arrived in a short period of time after political independence. What is more impressive is that the country had been able to continue this level of social development despite various downfalls in the economy and difficult political situations. In 2012 also, Sri Lanka's Human Development Index (HDI) stood at 0.715, ranking 92nd in the country list of 187 [41].

During the first three decades after independence (1948–1977), the government invested heavily in the social indicators. Since then the human development indicators have improved steadily but at a lesser momentum. Nevertheless in the entire post independent period when the economy grew on average by only 4.9 per cent per annum, the social attainments were far more commendable [35]. Amartya Sen argued that the goal of economic growth and increases in per capita incomes is to achieve human development. Therefore, if a country has achieved that objective even before attaining high levels of per capita income, as was the case of Sri Lanka, then it has achieved the ultimate objective of economic growth [35].

A Different View

Other economists have challenged this view and asserted that the country has low rates of economic growth owing to its social welfare policies. They have pointed out that the high achievements in human development came at a cost of economic growth and contend that the economic growth was retarded owing to the emphasis on welfare and ascribe the relatively slow economic growth, the persistence of poverty and unemployment to the country's high welfare expenditure. One of Sri Lanka's eminent economists, Lal Jayawardena put across the view that the country lost about one per cent of growth each year owing to welfare expenditure, particularly owing to the high cost of the rice subsidy. Jayawardena's argument was that if the country did not expend as much on the food subsidy, the investment of such funds would have resulted in substantial compounding of economic growth. We wonder whether the one per cent increase in growth would have so positively affected the quality of life for so many people, if the expenditure was not made.

Saman Kelegama contends that several reasons including the welfare-oriented policies accounted for retarding economic growth. He also identifies three important constraints apart from welfare policies as having retarded economic development: the ethnic conflict, security situation and the war. Furthermore, Kelegama contends that "Sri Lanka was also stricken by the doctrinaire of export pessimism and delayed moving towards an export-oriented strategy by overstaying as a closed economy for nearly two decades. When the economic liberalisation began in 1977, it was a latecomer to the industrial world and thus could not stage a significant breakthrough like the East Asian Tigers" [35].

Economic Indicators

GDP

The economy of Sri Lanka grew at an average rate of 5.33 per cent (table 1) per annum during 1999–2011. The Indian economy grew much faster at 7.28 per cent during the same period. The standard deviation is also lower for India (1.76) as compared to Sri Lanka (2.59). Thus the growth was not only higher but more stable for India. The paired t-test shows that the difference in growth rates between the two nations is statistically significant with p = 0.0009 and t statistics 3.952. India's growth has been statistically significantly higher.

It is reported that since around 1990, the rate of unemployment has been decreasing-from around 15-16 per cent in the early 1990s to 5.6 per cent in third quarter of 2007 [23]. The change in the sectoral composition of GDP shows the changing role of the agriculture sector in the economy, since 1960. By the end of 2006, the agriculture sector contributed the lowest percentage to GDP (16.8 per cent), industrial sector contributed 27 per cent of GDP and the services sector played the dominant role by contributing 56.2 per cent of the GDP (Central Bank of Sri Lanka, Annual Report, 2007).

Table 1: GDP - Real Growth Rate (Per cent)

Country	1999	2000	2001	2002	2003	2004	2005	2006	2007	2008	2009	2010	2011
Sri Lanka	3.7	5.6	-1	3.2	5.5	5.2	6	7.5	6.8	6	3.5	9.1	8.2
India	5.5	6	5.4	4.3	8.3	6.2	8.4	9.2	9	7.4	7.4	10.4	7.2

Source: *CIA World Factbook, 'GDP - real growth rate'.*

(http://www.indexmundi.com/g/g.aspx?c=ce&v=66)

and (http://www.indexmundi.com/g/g.aspx?c=in&v=66)

Mean (Sri Lanka) = 5.33

Mean (India) = 7.28

Standard deviation (Sri Lanka) = 2.59

Standard deviation (India) = 1.76

Paired t-test=0.000960594 and t statistics is 3.952

Inflation

The inflation rate for Sri Lanka (table 2) was an average 9.9 per cent between 1999 and 2011. For India, it was 6.5 per cent during the same period. The fluctuations in prices was also much more for Sri Lanka (standard deviation for inflation was 5.35) compared to India (2.66). In spite of such differences in mean and standard deviation, the paired t test shows that the statistical difference between the two series is not statistically significant. The p value is 0.4411. Over the years, both the nations have gone through a similar pattern of inflation changes.

Table 2: Inflation Rate (Consumer Prices)

Country	1999	2000	2001	2002	2003	2004	2005	2006	2007	2008	2009	2010	2011
Sri Lanka	6	8.5	14.2	9.6	6.3	5.8	11.6	12.1	15.8	22.6	3	5.6	7
India	6.7	5.4	3.7	5.4	3.8	4.2	4.2	5.3	6.4	8.3	10.9	11.7	8.9

Source: *CIA World Factbook, 'Inflation rate'.*

(http://www.indexmundi.com/g/g.aspx? v=71&c=ce&l=en–Sri Lanka and http://www.indexmundi.com/g/g.aspx?v=71&c=in&l=en- India)

Mean (Sri Lanka) = 9.9 and Standard deviation (Sri Lanka) = 5.35
Mean (India) = 6.5 and Standard deviation (India) = 2.66
Paired t-test= 0.044115966

Poverty

Although India has much higher economic growth rates, its poverty levels (21.9 per cent) are much higher than Sri Lanka (8.9 per cent). The Tables 3a and 3b shows how the Sri Lankan model of development with lesser growth in economic sectors and more in social sectors actually paid off.

Table 3a. Poverty Head Count Ratio (per cent of Population)

Country	1991	1996	2002	2007	2010
Sri Lanka	26.1%	28.8%	22.7%	15.2%	8.9%

Source: *Wright Thom, 'Grain Annual 2013', 2013.*

(http://gain.fas.usda.gov/Recent%20GAIN%20Publications Grain%20Annual%202013_New%20Delhi_Sri%20Lanka_5-6-2013.pdf)

Table 3b: Poverty Head Count Ratio (Per cent of Population)

Country	1994	2005	2010	2012
India	45.3%	37.2%	29.8%	21.9%

Source: *World Bank, 'World development indicators'.*

(http://data.worldbank.org/country/india)

Per Capita Food Availability

The per capita rice availability in Sri Lanka was 104.57 kg per capita per year in 2006 and 101.13 kg per capita per year in 2007 [1]. We do not have per capita rice availability figures for India-and it is not very relevant indicator for grain availability in India anyway-but the per capita food grain availability in India was 162.5 kg per capita per year in 2006 and 160.4 kg per capita per year in 2007 [33]. However given that 98 per cent of cereal production in Sri Lanka is of rice [15], the cereals comparison would not be too farfetched with just rice.

However, food grains include pulses too, so we can assume that the availability in the two nations is comparable for food.

Indicators of Development

Global Hunger Index:-

Sri Lanka ranked 43rd across 120 developing nations in 2013 Global Hunger Index figure. Compared to its South Asian neighbors (Table 4), this was a much better performance. Countries with GHI between 10.0 and 19.9 are classified as serious; Sri Lanka falls in this category.

India used to be in 'extremely alarming' category (GHI ≥ 30) in 1990, now it falls in 'alarming' (GHI between 20.0 and 29.9) category. India ranked 63rd in the Global Hunger Index, making a marginal improvement since 2012, but continues to languish far behind other emerging economies. The score for the country improved slightly from 22.9 in 2012 to 21.3 this year [38]. It has a long way to go to equal the achievements of Sri Lanka.

Table 4: Global Hunger Index

Country	1990	1995	2000	2005	2013
Sri Lanka	22.3	20.7	17.8	16.9	15.6
India	32.6	27.1	24.8	24.0	21.3

*(**Source:** IFPRI/ Concern/ Welthunger hilfe, 'Global hunger index'.)(http://en.wikipedia.org/wiki/Global_Hunger_Index)*

Colombo page, 'Hunger index'. (http://www.colombopage.com/archive_13B/Oct15_1381848707CH.php)

Human Development Index (HDI)

Sri Lanka's HDI value for 2012 is 0.715 (Table 5) in the high human development category-positioning the country at 92 out of 187 countries and territories. Between 1980 and 2012, Sri Lanka's HDI value increased from 0.557 to 0.715, an increase of 28 per cent or average annual increase of about 0.8 per cent. India is far behind with its HDI value 0.554 in 2012.

Table 5: HDI

Country	HDI value	HDI rank
Sri Lanka	0.715	92
Pakistan	0.515	146
India	0.554	136
South Asia	0.558	–
High HDI	0.758	–

*[**Source:** Undp.org, 'Human Development Report 2013'.*

(http://hdr.undp.org/sites/default/files/Country-Profiles/LKA.pdf)]

Literacy

Sri Lanka had very high literacy rates for many years now. While India improved its rates from 52 per cent in 1995 to 74 per cent in 2011 (Table 6a), Sri Lanka's literacy rates had been higher than 90 per cent throughout this period (Table 6b). However, since 1995, the improvement in literacy rates has been very slow because the government did not put in any investment in the sector.

Table 6a: Literacy Rate adult Total (Per cent of People Ages 15 and Above)

Country	1995	2001	2003	2006	2011
India	52	61	59.5	62.75	74.04

*[**Source:** CIA World Factbook, 'Literacy rate'.*

(http://www.indexmundi.com/g/g.aspx?v=39&c=in&l=en)]

Table 6 b: Literacy Rate Adult Total (Per cent of People Ages 15 and Above)

Country	1995	2001	2006	2008	2010
Sri Lanka	90.2	90.7	90.8	90.6	91.2

*[**Source:** CIA World Factbook, 'Literacy rate'. (http://www.indexmundi.com/g/g.aspx?v=39&c=ce&l=en)]*

Health Indicators

No statistical tests are warranted to see that the Sri Lanka's health indicators are much better than India's.

As table 7 shows, the IMR of Sri Lanka (9.47) is almost 5 times lesser than India's (46.07). MMR, which is even more amenable to improvement through better nutrition and health services, (shown in Tables 8a and 8b), is more than six times better for Sri Lanka (29) compared to India (190).

Table 7: Infant Mortality Rate

Country	2000	2001	2002	2003	2004	2005	2006	2007	2008	2009	2010	2011	2012
Sri Lanka	16.51	16.08	15.65	15.22	14.78	14.35	13.97	19.45	19.01	18.57	18.14	9.7	9.47
India	64.9	63.19	61.47	59.59	57.92	56.29	54.63	34.61	32.31	30.15	49.13	47.57	46.07

*[**Source:** CIA World Factbook, 'Infant mortality rate'. (http://www.indexmundi.com/g/g.aspx?c=ce&v=29) and (http://www.indexmundi.com/g/g.aspx?c=in&v=29]*

Table 8a : Maternal Mortality Rate

Country	2008	2010	2013
India	240	220	190

***Source:** World Bank, 'Maternal mortality ratio'. [(http://data.worldbank.org/indicator/SH.STA.MMRT)]*

Table 8 b : Maternal Mortality Rate

Country	2000	2005	2008	2010	2013
Sri Lanka	58	44	39	35	29

*[**Source:** World Bank, 'Maternal mortality ratio'. (http://data.worldbank.org/indicator/SH.STA.MMRT)*

CIA World Factbook, 'Maternal mortality ratio'.

(http://www.indexmundi.com/g/g.aspx?v=2223&c=ce&l=en)]

Economic Development Strategies of Sri Lanka

1948 to 1955:

The prices of main export crops, tea and rubber, were rising. With this, people's incomes were increasing and there was high liquidity in the economy. But in this phase of boom, investment was not considered as seriously as it should have been to ensure sustenance. Government efforts to lay a good infrastructure foundation were far from adequate. As a result, the productive capacity of the economy remained static.

Further, the government policies did not lay any foundation to diversify agriculture, which led the economy to depend on the three main primary exports of tea, rubber and coconut until very recently. Nor did the government identify the importance of reducing Sri Lanka's dependence on agriculture by diverting investment or developing the industrial sector. [39]

1956 to 1977:

By the middle of 1955, the export boom collapsed. The economy started going down. Food subsidies absorbed a major portion of the government budget while imports kept increasing at a relatively higher rate than exports. Something needed to be done to improve the balance of trade. Another turning point came about in 1960; the government pursued a policy to protect the domestic market, by restricting imports through elevating the tariff levels, imposing licensing quotas and banning selected imports which would harm the domestic producers. As such, the Government followed a policy of Import Substitution in Industry (ISI).

In 1965, the new government decided to follow import substitution in agriculture sector also. Hence, the attention was diverted into import substituting subsidiary crops for plantation. Even with these efforts the economic condition kept deteriorating and the terms of trade were more unfavourable for the country. At this time the government introduced the fixed exchange rate system. By 1970s, Sri Lanka had become a highly restricted and regulated closed economy [39].

1977-1990:

From 1978 to present, the period represents the history of open market reforms. The reforms liberalised trade and exchange control. Restrictions were relaxed for foreign travel and movement of capital and labour. This created an environment which was favourable to attract more Foreign Direct Investment. In keeping with its more market oriented policies, the government removed the domestic control of prices and subsidies were significantly reduced or cancelled altogether. Import and distribution of essential food was more of a government monopoly in past; it was opened to the private sector. The government also invested heavily into infrastructure. All these led to growth in imports and exports as well as expansion of the economic activity within the country [39].

Recent Development Strategies

During 1990s, the government continued to pursue liberalisation. While considering (both domestic and foreign) private sector as the engine of growth by pursuing market friendly policies, government intervention was limited to areas where the markets failed to function, welfare was not a priority anymore.

The coalition government which took office in 2004 realised that the social policies are essential for growth and economic stability. The policy of the new regime emphasised an equitable distribution of income by providing enhanced relief for the poor and vulnerable while encouraging more economic activities from the rural areas, through support of small and medium enterprises. Some polices such as restoring the fertiliser subsidy, broadening the poverty alleviation programme, continuation of electricity, transport and petroleum subsidies were carried out to address the issues of poverty and unbalanced growth at that time [41].

There are some concerns that the path, process and priorities of economic growth may be neglectful of the social welfare of the country. These concerns are not without a basis. The expenditure on education and health, as a percentage of GDP, is declining and woefully inadequate to sustain the achievements in these areas. Besides, there are new concerns in the post war era. The rehabilitation of the people whose livelihoods have been shattered by the war and the large number of orphans and displaced persons and persons physically maimed and in traumatic mental conditions that require medical attention and welfare measures poses serious challenges. And then there are the emerging problems of a rapidly ageing population [35].

Social Development

Introduction

The investment in human capital formation in Sri Lanka mainly starts with the free education policy and the free health policy (Jayaweera, 1998-[21]). It is important to note that the country complemented the human resource development policies by emphasising on protecting the minimum consumption levels. The proactive health and education services in combination had contributed to the impressive achievements in human and social development. The strong national economy of 1940s and 1950s supported such policies. With the economic liberalisation, the structural adjustment policies undermined the importance of investment in these sectors. Further the civil conflict in the country imposed heavy fiscal constraints on investment in human development.

The proportion of public expenditure on welfare during the time period 1958–68 amounted to about 40 per cent of the total public expenditure or 10-12 per cent of GDP. This figure includes the social welfare expenditures on education, health, transport, food subsidy and public welfare assistance (Jayasuriya, 2000[20]). The government expenditure on social services/ welfare now remains at around Six–Seven per cent of GDP [41].

Education

The Education Act of 1945 which is the predecessor of the Kannangara Report of 1943 recommended a system of universal and compulsory free education from the Kindergarten level to the University level. This radical approach provided a degree of social mobility. It became the single most important determinant of the social and political transformation of Sri Lankan society (Jayasuriya, 2000[20]).

Health

The success of the health sector can be attributed to two main policies, mainly free health care in the public sector and provision of services close to the patient since the 1930s. The health service includes health care infrastructures such as medical facilities, hospitals, health personnel, medical equipment and supplies. The health care system also addresses the problems associated with various communicable and non-communicable diseases and also includes a strong immunisation programme against diseases such as tuberculosis, tetanus, polio and measles. Moreover, child and maternal care as well as various nutrition programmes are also covered extensively by the health care system. Sri Lanka health indicators show the success of the past policies [41]. However, after about the 1960s, public sector spending on health was reduced due to resource limitation (de Silva, 2004[12]).

Table 9: Use of Maternal Health Services, by Wealth Quintiles.

	Poorest	Richest	Population average
Per cent who received tetanus toxoid immunisation	93.2	95.5	95.1
Per cent who received drugs to prevent malaria	34.9	12.5	23.8
Per cent who gave birth in an institution	91.4	99.2	96.7
Per cent who gave birth in private hospital	0.2	20.0	5.5
Per cent who received prenatal visits by midwife	75.6	80.7	82.9
Per cent who visited facility for prenatal care	93.1	94.1	93.9
%who receiving postnatal visits by midwife or medical officer	69.0	68.8	72.6

Notes: World Bank staff calculations using Sri Lanka demographic and Health Survey (2000) refer to maternal health services utilised during pregnancy by currently women aged 15-49 years with births within five years preceding the survey.

*[**Source:** Unrisd.org, 'Development Strategies, Welfare Regime and Poverty Reduction in Sri Lanka'.*
(http://www.unrisd.org/80256B3C005BCCF9/search/684D 9493 D25F9E C6C1257A E8005CAC3B? Open Document)]

The above table (Table 9) shows that there are not many disparities among the lowest and the top most wealth quintiles use of maternal health services. It also shows that the poor are using the government health services more. The scenario has changed in India, where most people prefer private institutions because of deteriorating quality of the health services provided in government hospitals.

The public health expenditure remained at approximately 1.5 per cent of GDP throughout 1990s. However, since 2003, there is an increasing trend in total public health expenditure. The increased military expenditure had curtailed other social expenditure programmes. As a result, the expenditure on health had also been low till 2000s [41].

Food Security

Food subsidies and Food supplementation programmes in past

Food Subsidies

The Food Commissioner's Department set up in 1942 was given the task of distributing 2.5 pounds of rice per week per person at subsidised rates. This scheme, though not specifically targeted to mothers and children, could be considered to have had an impact on the MCH status by ensuring availability of a proportion of the nutrient requirement to a given family, especially during the war years. The food subsidy scheme underwent many changes in subsequent decades and has minimal inputs at the present time.

Food Supplementation Programmes

One of the earliest food supplementation programmes was the milk feeding scheme aimed at improving the nutritional status of pregnant women and pre-school children, introduced in the mid-1940s, by the Food Commissioner's Department in collaboration with the Department of Medical Services. During the past four decades, the ongoing programme has been the Thriposha programme implemented through the health services, whereby a food supplement based on corn and soya is provided to pregnant and lactating women.

Food Stamp Scheme

The food stamp scheme, one of the novel schemes for distributing food that has been very much appreciated and discussed by the policy makers, has been the innovation of Sri Lanka. It was introduced in 1979. Its history holds lessons for all developing nations [22].

Initially, Sri Lanka, just like India and many other nations had a subsidised food distribution scheme. Eligibility was almost universal. Rice, the staple food for the entire population, and other major commodities including wheat flour, sugar, powdered milk, were subsidised. The change in this policy towards food stamps was made in three steps carried out over two years:

- The first step was a means test conducted in January 1978. It was used to restrict subsidised rice to families whose monthly income were less than Rs. 300, and resulted in the restriction of rice rations to only 50 per cent of the population. There was no change in the quantity of the free ration (one pound of rice per person per week) or in the amount that could be purchased at a subsidised price.
- The second step was the change from rations shop to food stamps in September1979. After much publicity, households were required to apply for food stamps through a declaration of incomes and household composition. Under the food stamp scheme, households with an income of less than Rs. 3,600 per year, with marginal adjustments for larger families, would be issued food stamps. Unused food stamps would be deposited in the Post Office Savings Bank.
- The third phase in the food subsidy reforms was the elimination of subsidies on food prices. Under the rationing scheme, and during the first phase of the new scheme, prices subsidies remained on rice, flour, sugar and infant milk foods. By the end of 1982, these subsidies were almost totally eliminated.

It eventually was phased out in the early 1990s because of rising costs and inaccurate targeting [14].

The Beneficiaries

The change from general food subsidy to the food stamp scheme reduced the number of households receiving food subsidies by about half. Although rice rations were limited to half of the population in 1978, price subsidies on a few other major food items allowed almost all households to receive some elements of the food subsidy. It was only when these price subsidies were eliminated by mid- 1980 that only half of the households remained as beneficiaries.

Criticism

An income criterion does not consider vulnerability caused by non-monetary factors like sickness. Thus the eligibility criteria may have eliminated households with incomes just over the income cut-off point but who were needy due to other problems.

Between 1978–79 (before the policy change) and 1981–82 (after it), nearly 75 per cent of the households either maintained or increased their per capita calorie consumption. However, per capita calorie consumption of the bottom 20 per cent declined by about eight per cent, from an already low 1,490 calories during 1978–79 to 1,368 calories during 1981–82. It appears that these households were not able to take advantage of the new income-earning opportunities created by the economic reforms during this period. It also appears that the food stamp scheme was not effective in helping the most vulnerable households.

The nominal value of food stamps remained unchanged for a long time. Prices were increasing during the same period. According to the research report 'The Food Stamp Scheme in Sri Lanka: Costs, Benefits and Options of Modification' by Neville Edirising, he [14] says, "According to one estimate, the real value of the food stamps was cut in half'. It is suspected that a large share of food stamp outlay "leaked" to upper-income households and thus reduced the cost-effectiveness of the income transfer.

The estimated impact of food stamps on the nutrition welfare of preschool children revealed that the benefits they receive depend on intra-household food allocation practices [14].

The Sri Lankan experience illustrated the difficulty at that time in effectively targeting the policy. Many of those making use of the stamps were not among the poor: it suffered from the fact that others had been accustomed to low-price food under the previous policy. The expense of allowing non-eligible consumers to take advantage of the stamps prevented the government from raising the value of the stamps along with inflation. As the purchasing power of the stamps dropped, the value of the programme declined. Participation rates among the targeted families were also unsatisfactory and the impact on reducing hunger was deemed to be marginal. More accurate targeting proved to be beyond the ability of the government (for political not technical reasons), but at least the cost was much less than the general food subsidy scheme that preceded it. In effect, Sri Lanka had provided a valuable example of the practical problems of running a food stamp scheme over a period of fifteen years [22].

Current Policies

Food Schemes

The Government of Sri Lanka's policy response to malnutrition consists of three broad strategies: direct food assistance programmes, poverty reduction programmes, and the Ministry of Health's provision of an integrated package of maternal and child health and nutrition services. Smaller programs focused on behaviour change communication also exist. Other programs are aimed at reducing micronutrient deficiencies and food fortification. The programmes are largely funded and implemented by the public sector, with the support of international agencies and domestic non-governmental organisations [45].

Most improvements gained during the past two decades in maternal and child nutrition can be attributed to the successful, country wide maternal and child health programme of the Ministry of Health, Sri Lanka. Other supporting programmes to combat deficiencies of individual micronutrients have also contributed significantly [11].

Currently the government spends around Rs. 4.5 billion per annum on direct nutrition specific programmes (Poshana Malla,

Thriposha, infant milk food subsidy, school nutrition food programme) and approximately Rs.100 billion has been spent on nutrition related programmes. The key national level Nutrition intervention programmes carried out by Ministries include: (i) The integrated MCH and nutrition programme, (ii) Food subsidies and poverty alleviation programmes, (iii) School Nutrition Programmes, (iv) Supplementary feeding programmes for mothers and children, (v) Micro-nutrient supplementation programmes (vi) Food fortification, (vii) Food security programmes and (viii) Provincial and community level nutrition programmes [32].

National Supplementary Feeding Programme (Thriposha Programme)

Thriposha programme was initiated in 1973 by the Ministry of Health, with the assistance of CARE, to combat the high incidence of child malnutrition, low birth weight and micronutrient deficiencies. It produces around 1.5 million packets of Thriposha every month and they are distributed among around 750,000 pregnant women, breast feeding mothers and underweight children. For the current production amount which is only 60 per cent of the requirement, around 10,000 tonnes of maize and 8,000 tonnes of soya beans are required annually as ingredients. Sri Lanka government spends around 1.25 billion per year for the production of Thriposha [36].

Target group

1. All pregnant mothers (throughout their pregnancy) and lactating mothers (for a period of six months after delivery).
2. Children above six months to five years of age who are underweight, loss weight for consecutive three months and hospitalised children who fall in these categories.

Some issues

- **Sharing**- the supplement may be consumed by the entire family rather than the target beneficiary.
- **Substitution**- when the supplement is given, the mother may reduce the food intake of the child resulting in no increase in nutrient intake [36].

Table 10: Ingredients Present in Thriposha

Ingredients	Per cent
Maize	66
Soya	30
Full cream powder milk	3
Vitamin premix	0.1
Mineral premix	0.9

*[**Source:** Wayamba University of Sri Lanka, Department of Applied Nutrition, 2008. (http://www.slideshare.net/damithagunawardane/nutritional-interventions)]*

In January 2006, the food supplement was being supplied through 815 clinics to approximately 580,000 beneficiaries, out of a target population of approximately 1.1 million children and approximately 53 per cent coverage (Ministry of Healthcare and Nutrition and World Food Programme 2006). However, although 90 per cent of pregnant and lactating mothers had received Thriposha, only 57.5 per cent consumed it daily, according to a study conducted in13 districts across the country.

The major reasons that the targeted recipients did not consume it daily were inadequate supply and the need to share it with the rest of the family (Department of Census and Statistics and UNICEF 2003, 2004). Thriposha's effectiveness therefore appears to be limited because the most vulnerable groups do not consume sufficient amounts, and it is often regarded as a food substitute rather than a supplement as intended. In addition, supply and distribution problems have also been observed [45].

Nutrition Programme

Nutrition Allowance Programme is implemented with the objective of enhancing the nutrition level of pregnant and lactating mothers of low income families. It was commenced in 1992. An allowance of Rs. 200 is given as an aid to obtain a nutritional meal to the lactating mothers in Samurdhi Beneficiary families. The amount is given in the form of stamp and it is issued monthly for the period of 12 months from the date of baby birth [34].

Nutrition food package programme started in 2007. It gives

monthly stamp of Rs. 500 to the pregnant and lactating mothers. Now this programme is implemented in all 212 divisions and under this, the beneficiary can obtain a "Nutrition Food package" worth of Rs.500/- for 20 months [34].

School Mid-Day Meal Programme

The programme started in the year 2006 to provide a daily meal to school children. Sri Lanka government has allocated Rs.3000 million in the year 2013 on free mid-day meals programme for school children in order to enhance the health and nutrition conditions of children. The programme has been implemented in 7625 selected schools to provide mid-day meals for 1,014,901 children studying in grade one-five classes [10 and 37].

Samurdhi Programme

Samurdhi programme has been the single largest welfare programme since its inception in 1995, with multiple objectives serving 2 million households in Sri Lanka. However, 80 per cent of its outlay is on a food stamp programme (Glinskaya, 2000 [18]) and therefore qualifies as the major food security programme in the country, although its stated objective is poverty reduction.

The programme intends to target the 20 per cent of the population which is below poverty line but in practice covers only 50 per cent of the country's total poor households.

This is seen as a problem from an administration point of view. In poor provinces, Samurdhi is less effective in reaching the poor than in richer provinces. Samurdhi programme's food stamps presently serve the 30 per cent of households just above the poverty line.

Some objectives-

(a) Broadening their opportunities for income enhancement and employment.

(b) Integrating them into economic and social development activities.

(c) Linking family level economic activities with community development projects at village, district, divisional and provincial levels.

(d) Mobilising their participation in the planning and management of projects and schemes for their upliftment. [18 and 27]

1. Poshana Malla-

The nutrition intervention package "Poshana malla and Kiri weeduruwa" was launched in March 2007 for highly vulnerable families in 10 selected districts of Sri Lanka.

It was a pilot project to support these families for enhancing the nutrition value of their diets. 45.4 per cent of Samurdhi beneficiaries received poshana malla.

All pregnant mothers and, lactating mothers having children less than one year of age were selected from low income families (less than Rs. 5000/- per month) as the beneficiaries of the pilot project.

Basket of foods containing identified dry commodities to the value of Rs. 500/- was issued to all pregnant mothers for a period of 18 months, i.e. six months before delivery and 12 months after the delivery during the breast feeding period.

A glass of milk to children within the age group of 2 to 5 years also was included in this intervention with a view to cover the most vulnerable growth period of children [25].

Of all pregnant mothers, 84.0 per cent received iron tablets of which 90.5 per cent took them daily while 76.0 per cent received Thriposha. 100 per cent women had received "poshana malla" in the Colombo Municipal Council area and in the rest of the district 91.3 per cent received iron tablets of which 90.5 per cent took them daily.

On the other hand, 85.7 per cent received Thriposha in rural areas and only 4.8 per cent women had received "poshana malla". Thus the coverage of poshana malla seems to be higher in urban areas while Thriposha and iron tablet programmes seem to have better coverage in rural areas [13].

Compared to Samurdhi, the amount of money spent on Poshana malla is much lesser but that is because Samurdhi covers the entire family and also invests in income generation while Poshana malla is limited to pregnant and lactating mothers and the infants [13]. A comparison of all feeding programs is given in table 21.

How much is the Per cent of total government expenditure on Samurdhi?

The government estimation for Samurdhi relief and Kerosene oil allowance is 9300 Million rupees for 2011. This amounts to four per cent of total government expenditure and only one per cent of the country's GDP [24].

Table 11 Ministry of Economic Development (Estimates 2012 in Rs Million)

Description	2010	2011 Revised Budget	2012 Estimate	2013 Projections	2014 Projections	2011-2014 Total
Samurdhi	9,241	9,300	9,300	9,300	9,300	37,200
Poshana Malla	388	550	600	750	850	2,750

*[**Source:** Ministry of Economic Development 'Estimates 2012'.*

(http://203.94.72.22/depts/nbd/budgetestimates/2012/headlist/105/Head_105.pdf)]

2. World Food Programme

The United Nations World Food Program (WFP) has agreed to provide Sri Lanka food items worth Rs. 4,524 million (US dollars 34.7 million) during the next two years.

The financial assistance from January 1, 2013 to December 2014 is expected to cover maternal and child nutrition, food for those whose work involves risk, lunch for school children, benefitting a total of 325,000 per year. This is being done under the Protracted Relief and Recovery Operation (PRRO) jointly implemented by both Sri Lanka Government and the WFP since year 2002 for providing food assistance to vulnerable people affected by 30 years of armed conflict [28].

3. North-East Irrigated Agriculture Project

The primary goal of Sri Lanka's North-East Irrigated Agriculture Project was to restore irrigation schemes and rural roads as a means of restoring food security for displaced communities.

The project was designed as a simple community-based operation. The project managed to ensure food security for about 33,250 conflict-affected people in Sri Lanka–including displaced persons returning to their homes–in 396 villages in the project region.

This was achieved by rehabilitating the productive capacity of about 370 small irrigation schemes, which brought 21,944 hectares of farmland back into cultivation.

Some Highlights

- Production of rice, a staple food, increased by 90,113 metric tons.
- Access to farms and markets improved. This benefited 83,250 households (or 341,000 persons) in 375 villages through the rehabilitation of nearly 1,294 km of rural roads.
- New access to clean water benefited 19,400 families through the construction of new wells.
- Through 317 women's rural development societies, the project supported almost 19,000 household-based livelihood support activities.
- Total project cost was US$ 131.6 million, including US$ 25.5 million from IDA, US$5.1 million from the Government of Sri Lanka and US$ 1 million from local communities.

Table 12: Summary of Direct Food Assistance and Supplementary Feeding Programmes

Program	Coverage	Target population	Time period	Cost
Thriposha	National	Pregnant and lactating mothers (till 6 months after delivery) and children aged 6 months to 5 years showing growth faltering.	From 1970 to date	Per capita cost in 2004=SLR 1,698 per year.
Fortified corn-soya blend, household food security	33 MOH divisions; 93,241 beneficiaries	Pregnant and lactating mothers and children when food security is low	2002-2006	Per capita cost=SLR 1,766 per year.
Poshana Malla(PM) and Glass of milk programme(GOMP)	113 divisional out of 303 secretariats for PM and 10 divisional out of 303 for GOMP	Pregnant and lactating mothers for PM and children aged 2-5 years for GOMP from low income households.	2006 to date	
School feeding program	6124 schools-GOSL 6125 schools-WFP	GOSL-All children from grades 1-3 and 1-9 for WFP.	1970s to date	
Preschool nutrition program	8 districts	Preschool children	2004-2006	Total cost=SLR 3,600,000;per capita cost= SLR 750

*(**Source:**http://siteresources.worldbank.org/SRILANKAEXTN/Resources/233046-1222978473355/ch4LKNutritionOct2008-2.pdf)*

Cash Vs Food Transfer

In 2006, the World Food Programme (WFP), Oxfam (GB) and the International Food Policy Research Institute (IFPRI), through a unique collaboration, implemented a Cash Transfer Pilot Project (CTPP) in Sri Lanka in the aftermath of the tsunami. The key objective was to compare the impact of cash and food transfers on beneficiary households' food and livelihood security and on the local economy. We have studied this experiment with interest because it is very relevant for the discussions in this book.

The Findings of the Study are as follows

- The study found that in areas where markets were functioning and accessible, cash transfer was more cost-effective and preferred by beneficiaries. In those areas where markets were less functional or accessible, food assistance was more cost-effective and preferred by beneficiaries. Thus the study observes that the appropriateness of cash programming depends on market access and functioning (whether they are competitive and integrated), and security. Food aid is more appropriate in contexts where markets are not working well, where security conditions impose higher market transactions costs for consumers, and in situations of high and unpredictable inflation.
- In the poorer, more remote and conflict-ridden communities in the east, the decline in per capita daily calorie intake was significantly steeper for cash-receiving households (a decline of 535 calories) than for food-receiving households (a decline of 290 calories), suggesting that cash transfers there had a net negative effect on household calorie intake.
- The cash disbursement system was more cost-efficient (five per cent cheaper) than the food system in all geographical areas. The lower cost of delivering cash was largely due to the existence of a well- functioning bank network compared to the relatively high-costs of moving food.
- The cost-effectiveness of food was higher in the east in areas where the security was deteriorating, markets and banks were difficult to get to, and where market prices were prone to higher fluctuations. Conversely, the cash was more cost-effective in areas with integrated and competitive markets, better bank access and lower market-price fluctuation.

The study concludes saying that opportunities exist for using both interventions in parallel or in a phased approach. In the immediate aftermath of a disaster, food intervention may be more appropriate. Cash-based interventions may be gradually introduced as markets recover and could potentially be used as an exit strategy [31].

How did Maternal Mortality Rate Come Down?

Sri Lanka's IMR and MMR are comparable to any middle income nation. It is interesting to know how it managed to decrease its rates so effectively.

Field Health Services

This initiative was taken in 1926 to provide field-based health services though a system of health units. Two important policy decisions were taken to improve the MCH services: Introduction of health education as a component of such services and the integration of MCH services with other activities carried out in health units, e.g., sanitation, prevention of communicable diseases.

Institutional Health Facilities

Establishment of general hospitals and rural hospitals started from 1940s. Several local authorities, especially the Municipalities, provided Maternity Homes where deliveries were attended to by trained midwives. Facilities available at the institutions were improved with access to specialised services in the higher level of hospitals, which served as referral centers.

Provision of Antenatal Care

During the late pre- and early post-independence periods, antenatal services through clinics held at government institutions continued along with the expansion of the health unit system to provide domiciliary services for pregnant women by trained field midwives and antenatal clinics held in 'health centers' in health units.

Improving Availability of Skilled Attendance at Birth

This was a key strategy adopted through both expansion of the health unit system which provided assistance from trained midwives in home-based deliveries and by increasing the availability of facilities for institutional deliveries [17].

Conclusion

Sri Lanka's success in achieving high rank in HDI comes from a large amount of GDP (almost 15-16 per cent) spent on social sectors in its early days. That foundation is bearing the fruits even today when the expenditure has come down to 5.8 per cent [2].

This shows that a society with firm foundations of literacy and health would maintain high performance levels in human development indicators for long. They would make their children go to school, the number of institution based deliveries would remain high and hence the IMR would be low. The high level of awareness among population would ensure higher life expectancy. Birth rates would come down to optimal levels automatically.

Sri Lanka also experimented with less costly disbursement strategies of food stamps and cash transfers thus making its expenditure more effective. The results are not conclusively in favour of such schemes but the experiments have taught us how and where we can use such schemes as options.

References

1. Agriculture and Environment Statistics Division, 'Per capita availabilities of major food item'. (http://www.statistics.gov.lk/agriculture/indices/PerCapita.html)
2. Central Bank of Sri Lanka, 'Annual Report (various issues)', (http://www.undg.org/docs/13045/Sri%20Lanka%20-%20UNDAF-Context-Final. pdf)
3. CIA World Factbook, 'GDP – real growth rate'. (http://www. indexmundi. com/g/g. aspx?c = ce & v = 66) and (http://www.indexmundi. com/g/g. aspx? c = in & v = 66)
4. CIA World Factbook, 'Infant mortality rate'. (http://www.indexmundi.com/g/g.aspx?c=ce&v=29)and(http://www. indexmundi. com/g/g. aspx? c = in & v = 29
5. CIA World Factbook, 'Inflation rate'. (http://www.indexmundi.com/g/g.aspx?v=71&c=ce&l=en - Sri Lanka and http://www.indexmundi.com/g/g.aspx?v=71&c=in&l=en - India)

6. CIA World Factbook, 'Literacy rate'. (http://www.indexmundi.com/g/g.aspx?v=39&c=ce&l=en)

7. CIA World Factbook, 'Literacy rate'. (http://www.indexmundi.com/g/g.aspx?v=39&c=in&l=en)

8. CIA World Factbook, 'Maternal mortality ratio'. (http://www.indexmundi.com/g/g.aspx?v=2223&c=ce&l=en)

9. Colombo page, 'Hunger index'. (http://www.colombopage.com/archive_13B/Oct15_1381848707CH.php)

10. Danuma, 'Sri Lanka starts mid-day meal for school children'. (http://www.danuma.lkindex.php? option = com_content & view = article&id =1058% 3ASri+Lanka+starts+mid-day + meal + for + school + children & catid = 40% 3Ass&Itemid = 191 &lang=en)

11. De Silva Angela, A.M.A.S.B. Mahamithawa and Chandrani Piyasena, 'Maternal and child nutrition: the Sri Lankan perspective', 2009. (http://www.icmr.nic.in/ijmr/2009/november/1121.pdf)

12. De Silva, A., 'Overview of the Health Sector. In S. Kelegama (ed.), *Economic Policy in SL: Issues and Debates* - A Festschrift in Honour of Gamini Corea. SAGE Publications, New Delhi. pp 426-444', 2004.

13. Dr. Jayatissa Renuka and Dr. Hossain Moazzem S.M., 'Nutrition and Food Security Assessment in Sri Lanka', 2009. (http://home.wfp.org/stellent/groups/public/documents/ena/wfp245828.pdf)

14. Edirisinghe Neville, 'The food stamp scheme in Sri Lanka: Costs, benefits and options for modification', 1987. (http://www.ifpri.org/sites/default/files/pubs/pubs/abstract/58/rr58.pdf)

15. FAO, 'Food insecurity in Sri Lankan scenario; causes and remedies', 2006. (http://www.sab.ac.lk/journal_agri/papers/agri_vol2_2_2006/agri_vol2_2_2006_artical3.pdf)

16. Fao-rap, 'Agricultural Statistics of Sri Lanka'. (http://faorap-apcas.org/srilanka/AgStat4web(SriLanka)/agstat.htm)

17. Fernando Dulitha, Jayatilleka Anoma and Karunaratna Vinitha, 'Pregnancy–reducing maternal deaths and disability in Sri Lanka: National strategies'. (http://bmb.oxfordjournals.org/content/67/1/85.full)

18. Glinskaya Elena, 'An Empirical Evaluation Samurdhi Programme, the World Bank', 2000. (http://siteresources.worldbank.org/INTDECINEQ/Resources/SamurdhiJune042003.pdf)

19. IFPRI/ Concern/ Welthungerhilfe, 'Global hunger index'. (http://en.wikipedia.org/wiki/Global_Hunger_Index)

20. Jayasuriya, L., 'Welfarism and Politics in Sri Lanka: *Experience of a Third World Welfare State.* The University of Western Australia, Perth', 2000.

21. Jayaweera, S., 'Education in Sri Lanka–Fifty Years Since Independence". In Indraratna A.D.V. de S. (ed.), Fifty Years of Sri Lanka's Independence: A Socio Economic Review. Sri Lanka Institute of Social and Economic Studies (SLISES), Colombo. pp 311-341', 1998.

22. Josling Tim, 'Global Food Stamps: An Idea Worth Considering?', 2011. (http://www10.iadb.org/intal/intalcdi/PE/2011/08841.pdf)

23. Lakshman W. D., 'UNDAF Sri Lanka 2012-17: The Country Context'. (http://www.undg.org/docs/13045/Sri%20Lanka%20-%20UNDAF-Context-Final.pdf)

24. Law and society trust, 'What when and how much is government spending?' (http://www.lawandsocietytrust.org/PDF/resource/Briefing_Paper_en.pdf)

25. Medical research institute Sri Lanka, 'Poshana Malla'. (http://www.mri.gov.lk/nutrition/index.php/component/phocadownload/category/12-downloads?download=99:samurdhi-2007&start=20)

26. Ministry of Economic Development, 'Estimates 2012'. (http://203.94.72.22/depts/nbd/budgetestimates/2012/headlist/105/Head_105.pdf)

27. Ministry of economic development, 'Samurdhi'. (http://med.gov.lk/english/? page_id=21)

28. Ministry of economic development, 'WFP gives Sri Lanka food worth million 4,524'. (http://med.gov. lk/english/? p=14626)

29. Ministry of Statistics and Programme Implementation (MOSPI), 'India population'. (http://www.tradingeconomics.com/india/population)

30. Ministry of Statistics and Programme Implementation (MOSPI), 'Sri Lanka population', (http://www.tradingeconomics.com/sri-lanka/population)

31. Mohiddin Lili (Oxfam GB), Sharma Manohar (IFPRI) and Haller Anette (WFP Rome), 'Comparing cash and food transfers: findings from a

pilot project in Sri Lanka'. (http://www.ennonline.net/fex/30/comparingcashandfoodtransfers)

32. National nutrition council, 'Multi-sector action plan for nutrition'. (http://www.nutrition.gov.lk/images/13-12-3-V19-Final-MsAPN.pdf)
33. Planning commission of India, 'Net availability of food grains (per annum) in India'. (planningcommission.gov.in/sectors/agri_html/selagri/T1.25.xls)
34. Samurdhi dept, 'Department of the commissioner general of Samurdhi'. (http://samurdhidept.gov.lk/nutrition.html)
35. Sanderatne Nimal, 'Economic growth and social development', 2011. (http://www.sundaytimes.lk/110717/Columns/eco.html)
36. Slideshare, 'Nutritional interventions'. (http://www.slideshare.net/damithagunawardane/nutritional-interventions)
37. Srilanka foundation, 'Schools mid-day meal programme'. (http://srilankafoundation.com/2013/01/rs-3000-million-allocates-on-school-mid-day-meals-programme/)
38. The Hindu, 'Global hunger index'. (http://www.thehindu.com/news/national/india-still-far-behind-in-the-global-hunger-index/article5234511.ece)
39. Tudawe Indra, 'Chronic Poverty and Development Policy in Sri Lanka: Overview Study'. (http://www.chronicpoverty.org/uploads/publication_files/WP09_Tudawe.pdf)
40. Undp.org, 'Human Development Report 2013'. (http://hdr.undp.org/sites/default/files/Country-Profiles/LKA.pdf)
41. Unrisd.org, 'Development Strategies, Welfare Regime and Poverty Reduction in Sri Lanka'. (http://www.unrisd.org/80256B3C005BCCF9/search/684D9493D25F9EC6C1257AE8005CAC3B?OpenDocument)
42. Wayamba University of Sri Lanka, Department of Applied Nutrition, 2008. (http://www.slideshare.net/damithagunawardane/nutritional-interventions)
43. World Bank Indicators Databank, by topic The World Bank, 'Demographics of India'. (http://en.wikipedia.org/wiki/Demographics_of_India)

44. World Bank, 'Maternal mortality ratio'. (http://data.worldbank.org/indicator/SH.STA.MMRT)

45. World Bank, 'Policy and programmatic responses to malnutrition: What is being done?'. (http://siteresources.worldbank.org/SRILANKAEXTN/Resources/233046-1222978473355/ch4 LKNutrition Oct 2008-2 .pdf)

46. World Bank, 'World development indicators'. (http://data. worldbank.org/country/india)

47. World Bank, 'World development indicators'. (http://data. worldbank.org/country/sri-lanka)

48. Wright Thom, 'Grain Annual 2013', 2013. (http://gain.fas.usda.gov/Recent% 20GAIN% 20Publications/Grain% 20 Annual% 202013_New % 20 Delhi_Sri % 20 Lanka_5-6-2013. pdf)

3

Indonesia

Compilation: Koosh Doctor

Introduction

With 237 million inhabitants, Indonesia is the world's fourth most populous country. The country rose to lower middle income status in 2009 and has experienced a gradual reduction in overall poverty, from 17 per cent in 2004 to 13 per cent in early 2010 (Rural Poverty Portal 2012). Indonesia's economic growth is expected to reach six per cent in 2012 and 6.4 per cent in 2013. The country's gross national income per capita has steadily risen from $ 2,200 in the year 2000 to $ 3,720 in 2009.

In spite of this progress, the problem of poverty persists. Poverty is increasingly concentrated in rural areas, where 70 per cent of the population lives. Compared with 9.9 per cent of urban populations, 16.6 per cent of rural people are poor (Rural Poverty Portal 2012). In addition, approximately half the population lives just above the poverty line set at 200,262 rupiahs per month (US$ 22). These "near poor" households are particularly vulnerable to shocks such as food price spikes and disease outbreak. Food insecurity and under-nutrition are persistent challenges, and the country's stunting levels are alarmingly high. In 2007, an estimated 7.7 million children under 5 (36.8 per cent) were stunted (2007). The stunting rate is higher than 30 per cent in most districts (ranging from 23-58 per cent). (Food security portal)

Food Security Indicators

GDP

As Table 1 shows, India's nominal GDP is much higher than that of Indonesia, at times almost thrice as much in most years. However, this is not a correct comparison as India is a much larger country compared to Indonesia.

According to the World Bank data [13], the per capita income of India was $1499 in years 2009–2013 whereas that of Indonesia was $ 3475 for the same period. Thus Indonesia's per capita income is almost 2.5 times as much. Indonesia is a middle income nation whereas India is still a low income country.

Table 1: GDP of Indonesia and India (Billion USD)

Year	2000	2001	2002	2003	2004	2005	2006	2007	2008	2009	2010	2011
Indonesia	165.02	160.45	195.66	234.77	256.84	285.87	364.57	432.22	510.25	539.58	709.19	846.34
India	476.60	493.95	523.97	618.36	721.59	834.22	949.12	1238.7	1224.1	1365.4	1780.5	1880.1

(Source: [13])

Between 2000–2013 the average rate of inflation in Indonesia was 7.6 and 6.89 for India. The standard deviation is similar 3.21 for Indonesia and 3.08 for India.

The paired t-test shows that the difference between the two series is statistically not significant (t statistic is an insignificant 0.124 and p is 0.31).

In Indonesia, the fluctuations came intermittently whereas in India, it grew steadily in percentage. But overall average remains almost the same. In short, both economies are facing similar inflation situation.

Table 2: Inflation Rates in Indonesia

Year	2000	2001	2002	2003	2004	2005	2006	2007	2008	2009	2010	2011	2012	2013
Indonesia	3.74	11.47	11.96	6.78	6.06	10.4	13.34	6.4	10.21	4.45	5.12	5.38	4.28	6.82
India	4.02	3.77	4.31	3.81	3.77	4.25	5.79	6.39	8.32	10.83	12.11	8.87	9.3	10.92

*[**Source:** Inflation EU, "CPI Historic Inflation Indonesia", "CPI Historic Inflation India", for years 2001 to 2013 (http://www.inflation.eu/inflation-rates/indonesia/historic-inflation/cpi-inflation-indonesia-2000.aspx, http://www.inflation.eu/inflation-rates/india/historic-inflation/cpi-inflation-india-2000.aspx)]*

Health and Social Indicators

Hunger and Poverty

Indonesia

Poverty mitigation efforts in Indonesia have shown meaningful progress, which has been in accordance with the MDGs as was demonstrated by the reduced proportion of people living under the national poverty line, i.e., from 15.10 per cent (1990) to 11.66 per cent (2012). The rate of GDP growth per worker strengthened from 3.52 per cent (1990) to 5.04 per cent (2011) [10]. Additionally, a reduction was observed in the proportion of people suffering hunger (Table 3) between 1989 and 2010 as the prevalence of under-five children with low weight went down from 31 per cent to 17.9 per cent

Table 3: Underweight Children Aged less than five Years (Per cent)

Year	Indonesia	India
1989	31.0	59.5 (1990)
1992	29.8	50.7 (1993)
1998	25.8	41.1 (1997)
1999	22.8	44.4
2005	24.4	43.5 (2006)
2010	17.9	42

*(**Source**: WHO Global Health Observatory Data Repository: Indonesia Statistics summary)*

India

India has been moderately successful in reducing poverty. The Poverty Headcount Ratio is estimated to reach 18.6 per cent by 2015 (CSO, MDG Report 2009). India is likely to miss the target by about 3.5 percentage points.

Thus eradicating hunger remains a key challenge. The proportion of underweight children was 59.5 in 1989 (Table 3). It came down to 42 in 2010. It is estimated that malnourishment could decline to 40 per cent by 2015. This would still be below the target of reducing malnourishment to 28.6 per cent [9].

40 per cent of pregnant and lactating women in Indonesia suffer from iron deficiency anemia. It is also responsible for 25 per cent of maternal deaths and associated with low birth weight. In India, it is reported to be as high as 87 per cent [5].

The IMR (Table 4) is almost double in India compared to Indonesia. In both the countries, the infant mortality rate has been almost halved between 1994 and 2012. We used the semi log regression to compute the rate of decline:

Log (IMR) = a + b (year) where b directly gives the growth rate.

In Indonesia according to this semi log regression, IMR declined at the rate of four per cent (R square = 0.99, t statistics = 94.22). In India the decline was slightly slower at 3.38 per cent (R square = 0.99, T statistics = 116.7). Thus the rate of fall has been the same, but given the high rates in India, it needs a much faster rate of decline.

Table 4 : Infant Mortality Rate

Year	Indonesia	India
1994	53	80
1995	51	78
1996	49	76
1997	47	73
1998	45	71
1999	43	69
2000	41	67

2001	40	64
2002	38	62
2003	36	60
2004	35	58
2005	34	56
2006	32	54
2007	31	52
2008	30	51
2009	29	49
2010	28	47
2011	27	45
2012	26	44

*(**Source**-World Bank: Infant Mortality Rate -World Development Indicators)*

The literacy rates (Table 4) are also much better in Indonesia: 93 vs. 73 of India in 2011. India has to make progress in its social sectors much faster, if it wants to catch up with the middle income South East Asian nations.

Table 5: Literacy Rates, Adult Total (Per cent of People Ages 15 and above)

Year	Indonesia	India
2006	92	62
2011	93	73

*(**Source:** World Bank: Literacy Rate - World Development Indicators)*

History of food security in Indonesia and the stability of rice prices before crisis [11]

The Historical Record on Food Security

Legacy of the Dutch, 1650–1940

SunanAmangkurat I (1645-1677) prohibited the export of rice from Java in 1655 in response to a severe drought that sent rice prices up by 300 per cent. For the next two centuries rice prices were very unstable around a steeply rising trend, and in 1847 appeared the first recorded imports of rice to Java, from Saigon.

Basic Dutch policy was to minimise controls, subject to broadly satisfactory welfare levels for producers and consumers, although the latter generally fared better.

In 1863, for instance, the import duty on rice was annulled following a bad harvest. Efforts were made to increase production to keep rice prices low, and when prices fell drastically in the 1880s as part of the worldwide overproduction of cereals, the Dutch response was to require that all government needs be supplied from domestic supplies. In 1911 poor crops and the approaching world war sent rice prices up again, and again exports were prohibited.

A long period of declining rice prices began in 1930 due to Asian overproduction and the world economic crisis. Other food prices fell in step with rice prices, and farmers could not pay their taxes. The limit to the functioning of the free market had been met.

In March 1933, the Government decided to intervene. It put an end to the free import of rice and restricted it by a system of licenses. This meant more than merely a checking of free importation; it signified the intent to work toward a system of self-supply with regard to rice.

Javanese rice which until then had been offered chiefly in local markets had to find its way to all the Outer Provinces. In the few rice-surplus areas of these provinces, such as Bali, Lombok and South Celebes, an inter-insular rice trade had to be started.

The prices at which the imported rice was sold to the public were controlled; if they appeared to be much higher than the c.i.f. value, the price level was reduced to reasonable proportions by the expedient of sending further supplies to the district concerned. Provision had to be made to ensure that stocks were not left over at the end of the period of scarcity, which might be used by speculators to repress the prices of the new harvest. Another fact noted at this time was also that the price of India's exported rice was lower than the Javanese. So import duties were raised on Indian rice to protect the Javanese rice growers and their higher prices. However, doubts were raised about the advisability of such actions.

Efforts by the New Republic, 1945–58

Due to continuing inflation in 1950 and 1951, rice rations were distributed in kind to civil servants and the military (and their families) to protect their real income. No longer was the government rice agency interested solely in avoiding high rice prices during scarcity and low prices during surpluses. It now had fixed distribution commitments that had to be honored, month in and month out. A government that cannot pay its civil servants and army will fall. First claim on foreign exchange for imports and on the rupiah budget for domestic purposes went to rice.

Perpetual shortages of foreign exchange to buy foreign rice frequently caused the government to turn to the countryside for increased output. Good farming techniques were demonstrated at critical locations in the countryside and were to spread gradually from there. Rice prices were stable from 1952 to 1954, and plans were made to eliminate imports in 1955 on the basis of the promising trends. But yields in 1955 were lower than in the previous years, the JUBM was caught without stocks when rice prices started to rise sharply, and the production programme fell apart in the scramble to arrange emergency imports. Massive imports that arrived in 1956 were used to push rice prices down. Prices declined throughout most of that year and even dropped during the three pre-harvest months in 1957. Imports continued for the next few years on a large scale: an average of 770,000 tons per year from 1956 to 1958 compared with only 225,000 tons from 1953 to 1955. And yet rice prices more than doubled from early 1957 to late 1958 as part of the inflation created by budget deficits.

Sukarno's Guided Economy, 1958–66

Governors of surplus regions kept prices extremely low to serve the interests of their urban consumers and to reduce budget demands for providing rations for their civil servants and military, while governors of deficit regions, mostly in the outer islands, found ways to tap local export earnings in order to import rice. This contradiction made rice trade volatile.

In a clash between using foreign exchange for fertiliser or for rice imports, rice in the short run always won over rice in the long run. As Indonesia's balance of payments deteriorated, the reality of the impact of monthly distribution requirements on Indonesia's rice policy was a mortgaging of the future for the present. As the domestic economy deteriorated under the brunt of exploding government deficits, spiralling inflation, and negative investment, the rice economy crumbled as well. To pick the worst years, rice production dropped by 13.6 per cent on Java from 1960 to 1964.

The production problems on Java in the early 1960s were caused jointly by declining area harvested and declining yields. This was mostly due to deterioration in the rather sophisticated irrigation network in Java. As the extent of controlled irrigation declined, so did the area successfully double cropped. Inadequate and uncertain water supplies led to lower yields.

The failure of the farmer to treat his rice intensified the foreign exchange demands of imported rice. From 1961 to 1963 over a million tons a year were imported, and then the foreign exchange simply ran dry. Imports the following three years averaged only 290,000 tons and rice prices spiralled out of control.

The Stabilisation Years of the New Order, 1966–70

By March of 1966,when leadership was transferred to General Suharto, an evaluation showed no rice in the warehouses of the food agency (then called BPUP), no foreign exchange in the treasury, and an inflation rate of 600 per cent per year. The first task was to find new supplies of rice. In the months December 1965 to March 1966, there was an acute shortage of rice, particularly for government employees and members of the Armed Forces. There were sufficient stocks in the free market, but depletion of government stocks led the authorities to reverse the earlier decision to stop imports.

Some Indonesian experts doubt whether Indonesia needed to import rice in the sense that domestic production is sufficient if properly distributed to meet reasonable minimum requirements of the population. Even these experts, however, agreed that imports of rice (on the order of $30 million) will be unavoidable because the Government is unable, organisationally and politically, to purchase from domestic sources the rice needed by the Armed Forces and for distribution in kind to government employees. To import rice for these purposes was easier (to the cities by ship from abroad than by land transport from the villages). Any attempts to do without rice imports would run into strong opposition from the politically powerful beneficiaries.

By early 1967 General Suharto emerged sufficiently powerful to set the country on a course of stabilisation. Large imports of food aid commodities, mostly rice and wheat flour, were arranged to keep rice prices under control. The inflation rate was reduced from 650 per cent in 1966 to 120 per cent in 1967.

There was a strong case for using foreign exchange to buy fertiliser rather than rice. There was indeed increasing recognition of the short-sightedness of a price policy which, by artificially keeping down the price of rice while allowing the price of imported fertiliser to rise through currency depreciation, made it uneconomic for farmers to buy fertiliser to expand rice production. But to tackle the problem by raising the price of rice was enormously difficult politically; and to reduce the price of fertiliser would require new subsidies in the teeth of the Government's resolve to abolish subsidies.

Until September [1967] it seemed doubtful whether the Government would be willing to court political trouble by raising the price of rice closer to the cost of imported rice or to the level at which it would pay farmers to buy fertiliser to produce more rice. The September crisis forced the Government's hand.

Indonesia's Food Logistics Agency, BULOG, was established in 1967. Its terms of reference at its founding were twofold: stabilisation of rice prices and provision of monthly rice rations to the military and civil service. By the late 1980s BULOG was still charged with its original role in rice markets but was also responsible for handling or monitoring sugar, wheat, corn, soybeans and soymeal, and a number of lesser commodities.

Surviving the World Food Crisis, 1970–1974

The First Five-Year Development Plan (REPELITA) was drafted in 1968 and inaugurated April 1, 1969. It was built around self-sufficiency in rice. The programme stressed getting profitable inputs, subsidised credit, and information out to the farmers. Fertiliser distribution was partially turned over to the private market, with a charge to sell for no higher than the ceiling price of Rp 26.6 per kilogram for both urea and triple super phosphate (TSP). The price required a subsidy to distributors of Rp 7-8 per kilogram (in 1971), which was covered from the Development Budget. From the very beginning of the Government's efforts to support incentive prices, a fertiliser subsidy was an integral component of the calculations.

The second innovation was to implement an effective floor price for stalk paddy. With a national fertiliser price ceiling established, it was possible to establish a national floor price. The government felt it could likewise commit itself to a nation-wide ceiling price for rice. Medium quality rice in urban markets was not to sell for more than Rs 50 per kilogram. Although this margin was very narrow in terms of prevailing interest rates, the private trade did seem to find it profitable to carry stocks in 1970 and 1971, partly perhaps because of increased supplies from the second harvest.

By mid-1972 the new programmes looked like major success stories. Rice production was exceeding the high targets set in REPELITA, BULOG was so successful it took over handling responsibilities for wheat flour and sugar, and the National Planning Agency and Ministry of Finance were trying to find alternative sources of revenue to take the place of food aid counterpart funds, which seemed about to disappear. Instead, the generally good weather from 1968 to 1971 ran dry (throughout most of Asia). In addition, BULOG moved quickly to improve its buying standards in order to reduce storage losses, and ended up buying very little rice in 1972. In a repeat of 1969 (and 1967), the dry season was poor, BULOG stocks ran out, and imports were suddenly hard to find. The government lost control of the rice situation and reverted to emergency imports as the solution. More than a million tons of very expensive rice poured into Indonesia from mid-1972 to mid-1973, before the world market for rice all but disappeared. A year earlier it had seemed that no imports at all might be needed. Urban rice prices went through the ceiling price of Rp 50 per kilogram in late 1972 and were more than double that level by mid-1973.

The Drive to Self-Sufficiency, 1975–1983

From 1975 to 1983 BULOG implemented the government's floor and ceiling price policy and delivered monthly rations to the Budget Groups without a hitch. Supporting the floor price received top priority as a way of stimulating domestic rice production, a crucial task because of the perceived unreliability of the world rice market. As world rice markets returned to normal in the late 1970s and Indonesia's foreign exchange reserves remained ample, BULOG turned increasingly to imports to meet rising demand in urban markets. Imports from 1977 to 1980 averaged nearly two million metric tons per year, or about one-fifth of the total amount of rice traded internationally.

Managing Self-Sufficiency, 1984–1989

Substantial surpluses emerged in 1985, BULOG's warehouses were still full from the record 1984 procurement, and support of the floor price was unsuccessful. Rice prices fell 20 to 30 per cent below the floor price in many areas. Farmers lost heavily. Once again, BULOG was unprepared for managing surplus stocks. BULOG's cost price became significantly above market prices because of high storage costs on a very large rice stock. BULOG's bankruptcy was almost inevitable.

On the other hand, prices in the world rice market also collapsed. Prices were over $400 per ton in 1981 but had fallen to below $200 per ton by mid-1985 for qualities similar to those produced and consumed in Indonesia. If world prices for these qualities had remained at $400 per ton, there would have been no "rice crisis" in Indonesia. BULOG's "surplus" stocks would have been a valuable export commodity, and prices for farmers would easily have been supported.

The success of its rice production programme actually created problems for Indonesia. The conversion in the early 1980's of Indonesia from an importer of one to two million metric tons of rice annually to a net exporter had a dramatic effect on the world rice market. Although all grain prices were low in the mid-1980s, rice prices were depressed even relative to long-run relationships with wheat or corn. This shift in relative prices was due largely to the withdrawal of Indonesia from regular import status. Attempts by Indonesia to solve its rice surplus problems by exporting sizeable quantities into those market conditions exacerbated this depressing effect.

Price policy was reconsidered and a new pricing model was proposed. The new pricing model was used to reduce incentives gradually to rice farmers. Rice production slowed its rapid expansion and rested at a plateau from 1986 to 1988 that left procurement sharply below the levels of the previous five years. BULOG's surplus stocks were exported and used for distributions to the Budget Groups.

But the stocks were not replenished. By late 1987 the agency was unable to inject enough rice into retail markets to maintain price stability. The shortages caught everyone by surprise. An Asia-wide drought, plus reduced stocks in the United States flipped the thin international market back to shortage. With a relatively short delay, Indonesian rice prices followed world prices up, ending 1988 at rough parity. However, because Indonesia had maintained its domestic rice prices well above those in world markets during the worst of the surplus in 1985 and 1986, its price increases were relatively smaller than those in the world market. Once again it seemed as though there was a longer run vision behind the stabilisation programme. BULOG activities were badly disrupted during the episode.

An excellent harvest arrived in 1989. Abundant rains late in 1988 and very substantial price incentives for farmers, by some measures the highest since the late 1960s, produced a bumper crop. BULOG was able to procure over 2.5 million tons of rice, thus replenishing its buffer stock and returning the overall rice economy to an equilibrium not seen since the early 1980s.

Costs and Benefits of Self-Sufficiency on Trend, 1990–1994

This review of Indonesia's early experience with price stabilisation programmes and implementation by BULOG until self-sufficiency in rice was achieved highlights two issues. First, there had been a continuing concern to use the interventions that stabilise rice prices to induce investment by the private sector rather than displace it. The stable price environment itself was one inducement, reducing uncertainty about the range of price volatility. The margins provided between the floor price in rural markets and the retail price in urban markets were also consciously designed to be profitable for competitive private traders to be able to handle most of the rice marketed. In a normal year, BULOG would procure and distribute less than 10 per cent of the rice produced and consumed in Indonesia.

Second, there was a continuing effort to keep the costs incurred by BULOG low to stabilise rice prices from exceeding the benefits that are generated by such stabilisation. Four dimensions to this issue were identified. First, at the most general level, there has been continuing concern over the size of the benefits to the economy and to the society of stabilising rice prices, and how these benefits change over time. Second, BULOG pursued approaches to cost control that involved improved management techniques in the storage, handling, and distribution of rice. Third, for more than a decade, analyses had been conducted on the extent of cost reductions that could be achieved by phasing out distribution of rice to civil servants and the military (the Budget Groups). Rice distributions to the Budget Groups totalled approximately 1.5 million metric tons per year, or about 60 to 75 per cent of BULOG's normal annual turnover. It had been difficult to phase out these distributions because the monthly turnover in BULOG's stocks lowered storage losses through better inventory management, the price paid by the Ministry of Finance for this rice was high enough to cover all of BULOG's operating costs and was thus a mechanism for funding the price stabilisation policy itself, and because the President remained (adamantly) convinced of the efficacy of the programme. Fourth, after the expensive episode of subsidising exports of rice in 1993, serious discussions began on the nature of cost savings that would result from redesigning the policy approach to stabilising rice prices.

Benefits and Costs of Price Stabilisation

Dawe (1995) constructed rough measures of the quantitative impact of BULOG's rice price stabilisation activities on the historical rate of economic growth in Indonesia. BULOG's role in the growth process declined in importance overtime. By the end of the fifth Five-Year Plan in 1994, stabilisation activities contributed only 0.19 percentage points a year to economic growth. The decline in benefits from stabilising rice prices occurred because the share of rice in the economy fell over time. The impact of rice price stabilisation on investment and economic growth thus declines at higher levels of per capita income. At the same time, it was found that the agency incurred very high costs for rice stabilisation.

Greater Flexibility through Self-Sufficiency

A strategy designed to reduce costs of price stabilisation while still generating most of the benefits had several dimensions. First, the amount of rice in storage, and the average period it is held, had to be reduced. Second, the country needed to resort more to international trade in rice by routinely using imports. Third, by using variable imports as the balance wheel, rice supplies and prices could be stabilised with smaller domestic buffer stocks. A more trade-oriented stabilisation strategy reversed the upward trend, seen since the early 1980s, in the cost of price stabilisation.

Beginning in the early 1990s, BULOG was allowed to define self-sufficiency on trend rather than in year-to-year terms, and this tight relationship began to give way to greater flexibility. The result was quite promising. BULOG's annual costs were reduced by about $70 million in the first half of the 1990s solely because of the added flexibility permitted by the new policy of self-sufficiency on trend. During the subsequent five years, 1989 to 1994, despite the greater volume of imports and significantly lower average levels of stocks, the coefficient of variation of retail rice prices in Jakarta and Surabaya averaged less than 10 per cent. Flexibility in rice trade thus contributed significantly to a more efficient and a more effective BULOG.

The Crisis Years

The very severe el Nino-caused drought that started early in1997, the financial crisis that became especially severe in Indonesia in early 1998, and the political crisis that has not been entirely resolved by mid-2002, each contributed in powerful ways to changing BULOG's ability to stabilise rice prices. After the agency lost control of domestic rice prices in mid-1998, strong pressures from domestic politicians and foreign donors combined to strip BULOG of is monopoly over international trade in rice. By late 1999, its mandate to stabilise rice prices was revoked.

Table 5 shows that the growth rates of food grains production in Indonesia have been very high. For example, for rice, the average rate of increase was 5.17 per cent per year between 2007–2010. In spite of such impressive increases in production, Indonesia has remained an importer of food grains (Tables 6 and 7). However, the flexibility in foreign trade of rice and other grains, together with availability of foreign exchange from other sources (like petroleum till now), has freed the rice economy from extreme volatilities.

Table 6: Agriculture Production of Food grains: Indonesia

Grains	2007	2008	2009	2010
Rice Production	57,157,400	60,251,100	64,398,900	66,469,400
Maize Production	13,287,500	16,323,900	17,629,700	18,327,600
Soybean Production	592,634	776,491	974,512	907,031

(Source: *http://faostat.fao.org/site/339/default.aspx)*

Table 7: Food Grain Imports: Indonesia (MT)

Grains	2005	2006	2007	2008
Rice Imports	188,945	456,100	1,406,278	288,369
Maize Imports	185,597	1,775,321	701,953	286,541
Soybean Imports	1,086,180	1,132,140	2,240,800	1,173,100

(Source: *http://faostat.fao.org/site/339/default.aspx)*

Table 8: Indonesia's Foodgrain Exports.

Agricultural Exports	2006	2007	2008	2009
Rice Exports	940.00	1,194	1,867	2,395
Wheat Exports	16,106	73,800	39,500	15,291
Maize Exports	28,074	101,740	107,001	62,575
Soybean Exports	4,633	1,872	1,025	446.00

(Source: *http://faostat.fao.org/site/339/default.aspx)*

Current Government Schemes, Expenditure Involved

Raskin

Introduction

Raskin has origins in the GOI's response to the Asian Financial Crisis in the late 1990s, making it the longest servingof the current household-targeted social assistance transfer programs. Originally called *Operasi Pasar Khusus,* Indonesia's subsidised rice programme first provided (in July 1998) food consumption assistance to households suffering the effects of the Asian Financial Crisis. Raskin rice stocks are acquired through the purchase of wholesale quantities from domestic producers at fixed prices by the state-owned Bureau of Logistics, BULOG, which then delivers agreed quotas of rice to over 50,000 regional distribution points where eligible households may purchase fixed quantities at below-market prices.

Table 9: Raskin at a Glance.

Official name:	Berasuntuk Rakyat Miskin (Raskin)
Program type:	Subsidised Rice Delivery
Program Type and Inaugural Year:	Permanent, Tax-Financed,1998
Coverage:	National (100 per cent provinces,100 per cent of districts)
Official number of target beneficiaries:	18.5 million Households
Official value of benefit (2009):	Rp 720,720 per household per year
Benefit value from household records:	Rp 125,250 per year
Public expenditure (2009)	Rp 13 billion (US$ 1.25 billion)
Administrative cost per official target beneficiary, budgeted (2009)	Rp 162.703 (US$ 16)

Percentage of poor households covered from household records	76
Key policy and executing agency:	Kemenkokesra
Key implementation agencies (role):	Bulog (1st stage) and local administrations (2nd stage)
Support operations partners (role)	BPS (targeting and eligibility)
Local government participation	Delivery to households, socialisation, monitoring and evaluation (M&E) activities

*(**Source:** Kemenkeu, Susenas and World Bank staff calculations, 2010)*

Financial Data

The mandated monthly allotments have fluctuated, but on average between 15-20 kilograms are given per household per month. Raskin has become the largest of the permanent household-targeted social assistance programs by expenditure and the second-largest by official targeted coverage. As food items make up more than two-thirds of overall expenditure on the average consumption bundle for poor households, and rice alone makes up approximately one-quarter of overall expenditure. Thus Raskin potentially delivers two significant benefits to beneficiary households. First, Raskin could bring to poor households approximately half of their monthly rice consumption needs at subsidised prices, leaving them additional income to spend on other budget items. Second, the Raskin programme could keep rice prices stable for poor households by facilitating purchases of significant allotments of rice at preannounced prices [4].

Central Government expenditures on Raskin totalled nearly Rp 13 trillion in 2009 (approx. US$ 1.4billion at 2009 nominal exchange rates).In 2012, the Government allocated a budget of Rp 15.7 trillion (USD $1.334 billion) to subsidise 3.41 million tons of rice to be distributed to 17.5 million households (Indonesian Financial Note and Revised Budget, 2012). Government budget in 2012 was USD $ 134.2 billion[4]. The budget (2011's *Nota KeuangandanRencana APBN)* indicates that the 2005 Raskin programme officially targeted 11.1 million households at a cost of Rp 397,830 per household and by 2009 official target was 18.5 million households at a cost of Rp 720,000 per household.

Table 10: Raskin Expenditure Summary, 2005-2009

	2005	2006	2007	2008	2009
Total Raskin (Nominal, Rp billion) Analytical Series:	6,400	5,300	6,600	12,100	13,000
Total Raskin (Constant 2009 prices, Rpbn)	11,227	8,149	9,121	14,153	14,043
Total Raskin (US, Rp million)	656	580	720	1,240	1,256
Share of total central government spending (Per cent)	1.8	1.2	1.3	1.7	2.1
Memo Items (all officially announced/budgeted):					
Target, beneficiary households (Millions)	11.1	12.7	16.7	19.1	18.5
Number of months of delivery	12	10	11	12	12
Monthly rice entitlement per household (Kg)	15	13	9	13	15
Value of annual benefit per household (Rp)	397,830	405,632	382,890	517,650	720,720

*(**Source:** Kemenkeu, Susenas and World Bank staff calculations, 2010)*

Target Population

Raskin suffers from both type I and type II errors. In other words, it does not include the ones who should be included and it includes many who should not be included. Raskin is intended for poor and near poor households, but the lists used to determine quotas are imperfect and exclude many poor households.

At the same time, coverage has been extended in many different ways to non-poor households. This is a major drawback of Raskin. Of equal importance is the manner with which the consultative meetings are held. The meetings often include only a few political officials or other authorities. The meeting activities often do not include verification of beneficiaries from the centrally-produced quota lists [4].

BLT Temporary Cash Transfer [16]

Introduction

The *Bantuan Langsung Tunai* (BLT) programme had a clear and modest objective: supplement consumption for poor households facing unprecedented price increases. In 2005 subsidy cuts raised household fuel prices by an average of over 125 per cent. BLT, a direct cash transfer in four instalments over one year, was funded from the implied budgetary savings from subsidy reductions. It was targeted to the poor households who were benefiting least from the old subsidy regime and most at risk from the negative impacts on consumption from price increases. A mostly-similar BLT was introduced again in 2008 when international crises in both financial markets and in food prices combined with another domestic reduction to fuel subsidies.

Though designed and deployed in less than five months, BLT reached households everywhere in Indonesia. Over 19 million households (more than a third of all households in Indonesia), spread across all provinces, in even the most remote and topographically challenging regions, received BLT in 2005. Transfers of Rp 300,000 were delivered. The post office (PT Pos) distributed benefits in every one of Indonesia's provinces. Three additional Rp 300,000 payments, spread over the following year, were made before the BLT window closed permanently. A total of Rp 1.2 million per household was transferred.

In 2008, there were approximately 600,000 fewer beneficiaries but every province continued to be served. The number of transfers was reduced to three, and the total household transfer was Rp 900,000. The BLT window was closed after nine months.

These transfers were more than enough to cover increased expenditure on fuels. Health service utilisation increased, especially for those with health insurance (including GOI-provided health service fee waivers). Neither nutrition nor consumption of harmful goods was affected by the receipt of BLT transfers.

BLT was designed as emergency income support; it was not designed to affect household behaviour or permanently lower the poverty rate. BLT transfers were modest (around 15 per cent of the average consumption budget of target household), temporary, and unconditional and the program's objective was to provide income support during a time of emergency. BLT could not and did not address either household behaviours or the correlates of poverty. Programmes introduced during the same era, like education and health fee waivers, public works, and conditional cash transfers are better suited for those objectives.

Table 10: BLT at a Glance for 2008-09

Official Name:	BantuanLangsungTunai (BLT)
Programme Type:	Unconditional Cash Transfer (UCT)
Programme Type and inaugural year (start/usage year)	Temporary, tax financed, 2005-2006 and 2008-2009
Coverage:	National (100% provinces, 100% districts)
Official number of beneficiaries (2008-2009)	18.5 million households
Official value of benefit (2008-2009):	Rp 100,000 per month for 9 months (Rp 900,000 in total)
Public expenditure (2008-2009):	Rp 18,966 billion (US$ 1.8 billion)
Administrative cost per recipient (2008-2009):	Approx. Rp 50,000 (USD 5)
Percentage of poor households covered, household records (2008-2009):	54
Local Government participation:	Socialisation, card distribution, monitoring and evaluation

Salient Features of BLT

As mentioned earlier, approximately 19 million households received BLT in 2005. Programme designers correctly predicted that simple, rapid and reliable delivery would be the key to reaching poor households when they were most vulnerable, so they agreed on a cash benefit delivered directly to beneficiaries through the national postal system (PT Pos). BLT went from initial design to deployment in less than five months.

Approximately 50 per cent of all households in the poorest quintile (according to expenditure) received BLT. BLT incidence and coverage makes it the most well-targeted of the current national social assistance schemes in Indonesia. BLT's lower inclusion and exclusion errors resulted in a higher percentage of total benefits being received by target households.

Cost Effectiveness

BLT's estimated administrative costs are low relative to other Indonesian and international social assistance programmes. Data from the 2008–09 BLT indicate average administrative costs per beneficiary of approximately Rp 50,000 (roughly US$ five at current nominal exchange rates) with an overall administrative overhead ratio of 5.2 per cent. These costs are low compared to more permanent cash transfer interventions in Indonesia. For example, the GOI's pilot conditional cash transfer (PKH) – which has comparable generosity to BLT – is estimated to have had an administrative overhead of 14 to 15 per cent in 2008 and 2009 and average administrative costs per beneficiary of around US$ 20. Two pilot social cash transfer programmes with smaller coverage and larger benefits for the severely disabled and abandoned elderly had estimated administrative overheads of 11 to 13 per cent in 2009, and higher per beneficiary administrative costs of around US$ 50 per year. BLT costs are also relatively low compared to UCTs in other countries. For example, an international survey of 16 cash and near cash programmes found that average administrative costs were around eight per cent [16].

Program KeluargaHarapan (PKH): Conditional Cash Transfer Programme [6,8]

Introduction

The household CCT programme, PKH, provides cash transfers to households. The programme is designed to support demand-side improvements in education and health services for poor families, so that children from chronically poor families can escape the intra-generational poverty trap.

The Specific Objectives of the PKH Programme are:

1. To improve recipients' socio-economic conditions;
2. To improve recipients' education levels;
3. To improve the health and nutrition status of pregnant women, postnatal women, and children under five years of age in recipient households; and
4. To improve recipients' access to and quality of education and health services.

The fund is not directly given to eligible households; rather, the households collect the funds at the local post office. Beneficiary households are determined by combining geographical and household level targeting. Locations are first selected on the basis of several criteria: high incidence of poverty, high incidence of malnutrition, low transition rate from primary to secondary school education, inadequate supply of health and education facilities, and approval from the local government for residents to participate in the CCT projects.

Eligible households must be classified as chronically poor (RumahTanggaSangatMiskin), and meet one of the following conditions: have a child aged six-fifteen years; have a child under 18 years who has not completed primary school; have a child aged 0-six years; or have a pregnant/lactating mother.

Cash transfers are made to households on the condition that certain health and education-related obligations are met. The Ministry of Social Affairs is the implementing agency and the post office manages the transfer of funds.

Table 12: Conditional Cash Transfer Eligibility

Sector	Indicators
Health	(1) Four prenatal care visits for pregnant women at health institutions (2) Taking iron tablets during pregnancy (3) Delivery assisted by a trained health professional (4) Two postnatal care visits (5) Complete immunisations (BCG (Tuberculosis), DPT (Diphtheria,

	Whooping Cough, Tetanus), polio, measles, and Hepatitis B and additional immunisations for children aged 0-11 and 12-59 months (6) Ensuring of monthly weight increases for infants (7) Monthly weighing of children under three and bi-annually for under-fives (8) Vitamin A twice a year for under-fives
Education	(9) Enrolment of all children aged 6 to 12 years in primary school (10) Minimum attendance rate of 85 per cent for all primary school-aged children (11) Enrolment of all children aged 13 to 15 years in junior high school; and (12) Minimum attendance rate of 85 per cent for all junior secondary school-aged children

PKH applies the traditional CCT design with quarterly cash transfers to individual poor households. Health facilities and schools regularly report non-use of these services to the kecamatan PKH management office. If a PKH recipient fails to comply with the conditions, after a few warnings, the cash transfers are terminated.

Table 13: PKH Cash Transfer Scenario to Poor Families.

Transfer Scenario	**Transfer Amount per Poor Family Per Year (Rp)**
Fixed transfer	200,000
Transfer for poor families who have:	
a. Children aged under six years	800,000
b. Pregnant/lactating mother	800,000
c. Primary school aged children	400,000
d. Junior high school aged children	800,000
Average transfer per poor family	1,390,000
Minimum transfer per poor family	600,000
Maximum transfer per poor family	2,200,000

*(**Source:** Tim Penyusun 2007 Note: 1 US$ Rp 12,000.)*

Unlike the previous UCT programme that was introduced in 2005–06, where cash transfers were received by the head of the household, usually male, PKH transfers funds to the woman in the household to ensure the programme's effectiveness in improving the quality of education and health. This could be the mother (wife) or one of the adult women in the household (grandmother, aunt, or elder sister),who takes care of the children in the family and whose name is written on the PKH member card. Thus the programme is designed to provide women in the community access to additional cash. In exceptional circumstances where a woman cannot be the recipient of the funds, the household has to complete a special form.

Planned to run until 2015, the programme aims to reach approximately 6.5 million chronically poor families. Every household will receive the funds for a six-year period [8].

Conclusion

The lessons that India can draw from Indonesia are numerous. Indonesia achieved much lower rates of malnutrition and better social and health indicators compared to India. Its growth surely seems to have contributed to this achievement. Ownership of natural resources like coal and petroleum helped a lot in the country's growth.

Raskin of Indonesia seems just like the PDS programme of India. Both seem to be suffering from both inclusion and exclusion error. But Raskin gets one billion USD for its implementation whereas India has devoted almost 20 million USD for reaching out to its population. India has almost six times more population than Indonesia but the allocations are almost 20 times more. Are the Indonesians doing this in more cost effective manner? This needs to be further investigated. The direct and indirect cash transfers, especially the former for its wide reach, have many lessons to hold for India. It is an alternative route to achieving food security for the poor and as is reported, is actually much more cost effective. Direct cash transfers are planned by Indian government also; the methods of reaching the remote areas and the poorest can be learnt from Indonesia. They have successfully carried out these programmes.

The conditional cash transfer programme is also very innovative. The progress and effectiveness needs to be further investigated to determine its suitability for India.

The economic history of Indonesia also demonstrates how difficult it is to try to stabilise prices. Achieving food security through other means seems to be a better approach.

References

1. CSO, Ministry of Statistics and Programme Implementation, Government of India, "Millenium Development Goals – India Country Report 2009". http: //planipolis.iiep.unesco. org/upload/India/India_ MDG_ report_ 2009. pdf
2. Food and Agriculture Organisation Statistics: Indonesia, 2012 http:// faostat.fao.org/site/339/default.aspx
3. Food Security Portal, "Indonesia": Facilitated by IFPRI http:// www.foodsecurityportal.org/indonesia
4. Kemenkeu, Susenas and World Bank staff calculations (2010): "Raskin Subsidised Rice Delivery: Social Assistance Programme and Public Expenditure Review 3" http://www-wds.worldbank.org/external/ default/WDSContentServer/WDSP/IB/2012/03/06/000333038 _2012030 6005448/Rendered/PDF/673080 WP00PUB L0 Background 0 Paper 0030.pdf
5. K. Kalaivani (2009): Prevalence and consequences of anemia in pregnancy, Indian Journal of Medical Resources 130, November 2009, pp 627-633 http://icmr.nic.in/ijmr/2009/november/1125.pdf
6. Stella A. Hutagalung, Sirojuddin Arif and Widjajanti I. Suharyo (2009), "Problems and Challenges for the Indonesian Conditional Cash Transfer Programme-Program KeluargaHarapan (PKH)", Social Protection in Asia Working Paper Issue 04, May 2009 http:// www.academia.edu/1526387 Problems_ and_ Challenges_ for_the_ Indonesian_Conditional-Cash_Transfer_Programme_- _Program_Keluarga_Harapan_PKH_
7. Rural Poverty Portal, "Rural Poverty in Indonesia", powered by IFAD http://www.ruralpovertyportal.org/country/home/tags/indonesia
8. Tim Penyusun, LintasKementriandanLembaga (2007): "PedomanUmum PKH-ProgrammeKeluargaHarapan [General Guidelines of Household Conditional Cash Transfer 2007]", Jakarta

9. UNDP: Millenium Development Goals, "Eradicate Extreme Poverty and Hunger: Where do we stand?" (http://www.undp.org/content/undp/en/home/mdgoverview/mdg_goals/mdg1)

10. UNDP (2011), "Report on achievement of millennium development goals Indonesia", 2011. http://www.undp.or.id/pubs/docs/Report Per cent 20on Per cent 20the %20 Achievement% 20of% 20the% 20MDGs% 20in% 20Indonesia% 202011.pdf

11. WayanRusastra, I, Thompson, Geoff, Bottema, J.W. Taco, Baldwin, Robert (2008): "Food Security and Poverty in an Era of Decentralisation in Indonesia", CAPSA Working Paper 102, http://purl.umn.edu/96657

12. WHO: Global Health Observatory Data Repository, Indonesia Statistics Summary http://apps.who.int/gho/data/node.country.country-IDN?lang=en

13. World Bank: Gross Domestic Product -World Development Indicators, http://data.worldbank.org/indicator/NY.GDP.PCAP.CD

14. World Bank: Infant Mortality Rate -World Development Indicatorshttp://data.worldbank.org/indicator/SP.DYN.IMRT.IN

15. World Bank: Literacy Rate - World Development Indicators, http://data.worldbank.org/indicator/SE.ADT.LITR.ZS

16. World Bank Report (2012): BLT Temporary Unconditional Cash Transfer- Social Assistance Program and Public Expenditure Review 2http://www-wds.worldbank.org/external/defaultWDS ContentServer/WDSP/IB/2012/03/06/000333037_20120306001405/Rendered/ PDF/673240 WP0BLT0T00 Box 367866 B00 PUBLIC0. pdf

4

China

Compilation: Sanchit Agarwal

Introduction [1]

China is the third largest country in the world in terms of geographical area but it houses the world's largest population. The country is a developing nation with a capitalist economy that is controlled politically by communist leadership. Chinese civilisation began more than 5,000 years ago and the nation has played a crucial role in world history and is continuing to do so today.

Government of China

China's government is a communist state with unicameral legislative branch called the National People's Congress that is made up of 2,987 members from the municipal, regional and provincial level. There is also a judicial branch comprised of the Supreme People's Court, Local People's Courts and Special People's Courts. China is divided into 23 provinces, five autonomous regions and four municipalities. National suffrage is 18 years of age and the main political party in China is the Chinese Communist Party (CCP). There are also smaller political parties in China, but all are controlled by the CCP.

Economics and Industry in China

China's economy has changed rapidly in recent decades. In the past, it was focused around a highly planned economic system with specialised communes and was closed to international trade and foreign relations. In the 1970s however, this began to change and today China is strongly economically tied to the world's countries. In 2008, China became the world's second largest economy.

Geography and Climate of China

China is located in Eastern Asia with its borders along several countries and the East China Sea, Korea Bay, the Yellow Sea, and the South China Sea. China is divided into three geographic regions: the mountains to the west, the various deserts and basins in the northeast and the low lying valleys and plains in the east. Most of China however consists of mountains and plateaus such as the Tibetan Plateau which leads into the Himalayan Mountains and Mount Everest. Because of its area and variations in topography, China's climate is also varied. In the south it is tropical, while the east is temperate and the Tibetan Plateau is cold and arid. The northern deserts are also arid and the northeast is cold temperate.

More Facts about China

- China instituted a One Child Policy in 1979 to control its growing population
- The majority of Chinese are non-denominational in religion, but 10 per cent are Buddhist
- China's population is expected to peak in 2026 at 1.4 billion. India will surpass China as the world's most populous country in 2025.

Indicators of Food Security

Demographics

Population	1,349,585,838 (July 2013 est.)
Age structure	**0-14 years:** 17.2% (male 124,773,577/female 107,286,198) **15-24 years:** 15.4% (male 109,922,192/female 98,325,568) **25-54 years:** 46.7% (male 322,161,347/female 308,101,780) **55-64 years:** 11.3% (male 77,374,476/female 75,289,733) **65 years and over:** 9.4% (male 60,597,243/female 65,753,724) (2013 est.)
Dependency ratios	**total dependency ratio:** 36.8 % **youth dependency ratio:** 24.7 % **elderly dependency ratio:** 12.1 % **potential support ratio:** 8.2 (2013)
Median age	**total:** 36.3 years **male:** 35.5 years **female:** 37.2 years (2013 est.)
Population growth rate	0.46% (2013 est.)

[Source: CIA World Factbook, 2011(http://www.indexmundi.com/china/demographics_profile.html)]

(http://www.ifpri.org/book-8018/ourwork/researcharea/global-hunger-index)

Prevalence of Undernourished Population (Per cent)

GDP Growth (Real)

China's has been a phenomenal growth story. Since last few decades, it was growing at one of the highest growth rates that any nation has seen. As can be seen from Table 1, the Chinese growth rates from 1999–2011 have been consistently higher than India's. The mean growth rate for the decade works out to be 9.07 for China and 7.44 for India. The growth has also been slightly steadier for China: the standard deviation is 1.29 for the Chinese series whereas it was 1.73 for India.

The results of paired t-test show that the Chinese growth rates are statistically significantly higher than those of India, with t test significant at 4.72 and the p value for one tail at 0.0003.

Table 1: GDP Growth (Real) of China and India.

Country	1999	2000	2002	2003	2004	2005	2006	2007	2008	2009	2010	2011	Avg	STD
China	7	8	8	8	9.1	9.1	10.2	11.9	9	9.1	10.3	9.2	9.075	1.29
India	5.5	6	4.3	8.3	6.2	8.4	9.2	9	7.4	7.4	10.4	7.2	7.44	1.73

*[**Source:** CIA World Factbook, 2011(http://www.indexmundi.com/g/g.aspx? v=66 &c=ch&c=in&l=en)]*

Inflation Rate

Although the GDP was growing at consistently higher rates in China, the impact of such overheated economy has not been felt much on the prices, as can be seen from the Table 2 below. The highest inflation that they experienced in last decade has been 5.9 per cent while India experienced inflation in double digits in some years. Stable prices contribute a lot to economic security including food security because prices of basic commodities do not go beyond the means of the poor. High growth rates, accompanied by reasonably stable prices, seem to be some of the important factors in causing a decline in poverty and hunger.

Table 2: Inflation Rate of China and India

Country	1999	2000	2002	2003	2004	2005	2006	2007	2008	2009	2010	2011	Avg	STD
China	-1.3	0.4	-0.8	1.2	4.1	1.8	1.5	4.8	5.9	-0.7	5	5.5	2.28	2.65
India	6.7	5.4	5.4	3.8	4.2	4.2	5.3	6.4	8.3	10.9	117.7	8.9	6.76	2.63

*[**Source**: CIA World Factbook, 2011 (http://www.indexmundi.com/g/g.aspx?v =71&c =ch&c=i n&l=en)]*

Indicators of Malnutrition

China's infant mortality rates are much below those of India (Table 3). In 2011, India's infant mortality rate was three times higher than that of China. Thus India still remains far below the MDG target, whereas China has already attained it.

However, life expectancy is only slightly higher (Table 7) for China. The paired t-test for the life expectancy series shows the difference in the two series to be statistically insignificant. The t- test is insignificant at 1.08 and p value at 0.152. Thus the life expectancy series is statistically not very different in the two nations.

Table 3: Infant Mortality Rate in China and India

Year	2000	2001	2002	2003	2004	2005	2006	2007	2008	2009	2010	2011
China	28.92	28.02	27.25	25.26	25.28	24.18	23.12	22.12	21.16	20.25	16.15	16.06
India	64.9	63.19	61.47	59.59	57.92	56.29	54.63	34.61	32.31	30.15	49.13	47.57

*[**Source:** CIA World Factbook, 2011 (http://www.indexmundi.com/g/g.aspx?v = 29&c=ch&c= in&l= en)]*

Table 4: Life Expectancy at Birth in China and India

Country	2000	2001	2002	2003	2004	2005	2006	2007	2008	2009	2010	2011
China	71.38	71.62	710.86	72.22	71.96	72.27	72.58	72.88	73.18	73.47	74.51	74.68
India	62.5	62.86	63.23	63.62	63.99	94.35	64.71	68.59	69.25	69.89	66.46	66.8

*[**Source:** CIA World Factbook, 2011(http://www.indexmundi.com/g/g.aspx v =30&c=ch&c = in&l = en)]*

Global hunger index, given in Table 5, shows that China has made remarkable progress in tackling the problem of hunger. India's indicator of the hungry is almost four times higher than China's. Just like Infant mortality, even in case of underweight children (Table 6), China has been far ahead of India: only three per cent of the children in China are underweight as opposed to almost 40 per cent in India. India will take a long time to reach the progress that China has achieved.

Table 5: Global Hunger Index of China and India

Country	1990	1995	2000	2005	2013
China	13.0	10.4	8.4	6.7	5.5
India	32.6	27.1	24.8	24.0	21.3

*[**Source**: Global Hunger Index Data from IFPRI (http://www.ifpri.org/book-8018/ourwork/researcharea/global-hunger-index)]*

Table 6: Prevalence of Underweight Children Under five Year (Per cent) in China and India

Country	1988-92	1993-97	1998-02	2003-07	2008-12
China	12.6	10.7	7.4	4.5	3.4
India	59.5	45.9	44.4	43.5	40.2

*(**Source:** Global Hunger Index Data from IFPRI)*

Human Development index, which looks at the general well-being of a population by looking at income, education and health, also puts China above India. The series given in Table 7 shows that China improved in its position from. 40 points in 1980 to 0.69 points in 2011.

On the other hand, India advanced from 0.34 points in 1980 to 0.55 points in 2011. The paired t-test shows that the difference in the two series is statistically significant: t-test was strongly significant at 8.29 and p value for the two tails was 0.0002. Thus China's human development index figures were statistically significantly better than India's.

The literacy rates are also much higher for China (90.94 per cent) compared to India (74.04 per cent) in 2010-11. Thus on economic, social and health fronts, China is far ahead of India.

Table 7 : Human Development Index of China and India.

Country	1980	1990	2000	2005	2009	2010	2011
China	0.40	0.45	0.49	0.54	0.59	0.63	0.69
India	0.34	0.38	0.41	0.44	0.46	0.50	0.55

*(**Source** : UNDP, "Human Development Report" of various years)*

Table 8: Literacy Rate in China and India

Country	1980	1990	2000	2010
China	51.14	68.07	86.53	90.94
India	43.06	52.2	65.01 (2001)	74.04 (2011)

*[**Source** : Trading Economics. 2014(http://www.tradingeconomics.com/china/literacy-rate-adult-female-per cent-of-females-ages-15-and-above-wb-data.html)*

CIA World Factbook, 2011 (http://www.indexmundi.com/facts/china/literacy-rate) Census of India, Chapter 6, Status of Literacy, 2011.(http://censusindia.gov.in/2011-provresults/data_files/mp/07Literacy.pdf)]

Cereal Yield (kg Per Hectare)

China has higher population than India but it also has higher yields per hectare. The yields are almost double than those of India.

The average rate of growth of yields in China, given by the β of semi log regression equation, Log (yields) = a + b (year) is 2.4 per cent p.a. (R^2= 0.81 and t statistic was significant at 6.21).

A similar regression analysis for India yields the b to be 1.9 per cent p.a. (R^2 = 0.67 and t statistics was significant at 4.31.

In other words, not only are the yields higher in China but they have been growing at a faster rate compared to India. In many ways, this addresses the supply side of the food security problem. Food availability in the country rises because of such large production. If most rural families have land, their food security automatically gets taken care of with such large yields.

Needless to say, India has a lot to learn yet in its agriculture sector.

Table 9 : Cereal Yield in China and India

2011	2010	2009	2008	2007	2006	2005	2004	2003	2002	2001	Year
5837.4	5520.6	5520.6	5549.7	5319.8	5313.4	5225.5	5189.7	4877.7	4889.7	4208.2	China
2953.6	2536.6	2571.9	2672.6	2583.3	2446.5	2411.6	2350	2399.4	2187.3	2423.1	India

*[**Source** : Trading Economics, 2014 (http://www.tradingeconomics.com/china/cereal-yield-kg-per-hectare-wb-data.html)]*

Economic History/ Food Security History

History of PDS in China [7]

When the Communist Party of China came to power in 1949, there was an overall food shortage in the country as a consequence of decades of wars. The new government took various measures in order to promote grain production, crack down on hoarding and speculation, and establish and strengthen state grain organisations. By the end of 1950, the grain situation was basically brought under control and state grain organisations established a commanding position in the grain market.

China started its First Five-Year Plan in 1953. With economic reconstruction underway on a large scale, the demand for grain outpaced availability. In October 1953, it was proposed that the government should procure grain directly from rural areas and supply it to consumers in urban areas through a ration system. This scheme was endorsed by the government and implemented from December 1953. Under this scheme, state grain agencies would be the sole buyers and sole sellers in the grain market. This scheme was called the "Unified Grain Procurement and Sale System", and the sale part, the "Unified Grain Sale System" (UGSS).

The UGSS covered three kinds of buyers: (1) the non-agricultural population (urban) who were issued with grain coupons, (2) the agricultural population who were engaged in non-grain businesses or did not produce enough grain, (3) other grain users (e.g., restaurants, bakeries, and food-processing factories or factories using grain as inputs).

The grain coupons were an important element of the rationing system. They could be used in government grain stores, restaurants, and manufactured food stores, etc. Usually they could only be used within the issuing area (e.g., a city or a province) but a local grain coupon could be exchanged for a more general one (i.e., issued by a higher level government) to facilitate travelling beyond the local area. The local grain coupons were usually issued monthly, could be used any time or sometime for a specific period, and the proportion of the fine to the coarse grains was often fixed. The practices varied somewhat across provinces/cities. Food items sold through government grain shops primarily included cereals (chiefly rice and wheat flour), other coarse grains, and edible oil.

To the early 1990s, the system underwent few significant changes. A few minor changes include the following. (1) Per capita ration was reduced by one kilogram per month in late1960 in response to the nation-wide famine. (2) There were three sale price increases in the mid-1960s. (3) In 1985, the sale price of grain supplied to those qualified agricultural population was increased to equal the procurement price (the non-agricultural population was still provided with grains at the unified sale price which was below its procurement price). In the same year, changes in the provision to other grain users were also made.

Starting from the early 1980s, the subsidy on grain consumption increased quickly and by1990 it reached a total of 27 billion yuan as a result of the significant increase in the procurement price but no increase in the grain sale price to the non-agricultural population. This soon drew much attention and debate in the country. Some advocated that the government should substantially reduce its control over the grain marketing business and that grain prices should be left to be determined by the market (see, for example, Liu, Zhou, Gao and Chen 1986; Cheng, Lu, and Yan 1987; Yu 1987). But others argued that China's grain cannot be totally left to the market because of the critical importance of grain in feeding the people and maintaining social stability (Liu 1986; Ma 1987). The government left the sale price of grain untouched in order to maintain social stability.

In May 1991, the government moved to reduce subsidies for rationed grain by increasing the unified grain sale price. After this increase, however, the sale price was still lower than the procurement price. In April 1992, the sale price was further increased to equal the grain procurement price.

Market grain prices were low as a result of good harvests since 1990. They became not much different from prices of grains from government shops. Consumers bought more grains from the market. This enabled the reform trials of the UGSS to speed up and to be applied on a wider scale, particularly during late 1992 and early 1993. By about mid- 1993, the UGSS, which was in use for about four decades, disappeared in most areas of the country.

From October 1993, the grain price in the free market increased sharply with the situation aggravated by panic buying. Having been so sensitive about the grain price, the government immediately mobilised various resources to cope with this price surge, including placing a ceiling price on grain traded in the free market. Grain price was brought under control by about early December of that year with heavy administrative intervention.

In the first half of 1994, price fluctuations continued in some areas. From July 1994, grain prices again rose quickly all over the country. From late 1994, some local governments restored the rationing methods to supply grain to urban consumers at a subsidised rate.

During 1995, there were no worrying price surges, thanks to increased grain imports by the government and increased grain supply through government shops at subsidised prices. Some areas started to reintroduce coupons and by September some half of the provinces restored the use of coupons (Anon. 1995; Ka 1995).

Following the abolishment of the UGSS, local governments are assigned the primary responsibility of handling grain matters under their jurisdiction. Consequently, although they still procure grains under a quota regime at a government-set price, the methods of public distribution of grains are not identical across regions. Some cities sell subsidised grains through government shops without a ration; others do so with a ration. But few of them used any targeting, except in a few cities, e.g., Shanghai and Beijing, where attempts have been made to targeting the low-income population.

Agricultural Policy Reforms to Bring about Food Security by Increasing Production [4]

In 1976, when China was still under the planned economy, the nationwide average annual per capita income in the People's Communes (the agricultural collectives) amounted to 60.2 Yuan, which was lower than that of 1956 at constant prices. At that time, more than one third of farmers' households were in debt, and about 100 million farming people suffered from a shortage of food. By 1978, China was no longer self-sufficient in grain and had to rely on imports to supply 40 per cent of the urban population.

Farmers created a responsibility system with an output contract for the work groups in 1978 and developed it into a household contract system in 1981. The initiative of the farmers was given a positive response by top-level leadership, and eventually, it led to the abolition of the People's Communes. The following principle measures in the reform programme were implemented during the 1980s:

- A family farm system replaced the collective system. Every peasant household now ran a small farm with the right of usufruct for an average 0.5 hectares of farmland.
- Price and marketing policies changed by means of lifting the state monopoly on purchasing and marketing farm products, opening the free market, introducing the state contract purchase system and raising prices.
- Structural policies were devised to stimulate agricultural diversification.
- Rural non-agricultural economic development policies were designed in order to shift surplus labour away from agriculture.

The above-listed measures were issued through Document No.1 of the Central Committee of the Chinese Communist Party at the beginning of each year during 1982–1986, which indicates the first priority in the decision-making fields of top-level leadership. Document No.1 first plays a role in getting the local governments to pay special attention to food production and rural development. Second, it publicises the measures to target the short-term goals of the policies for the year that the document refers to. Third, it reasserts the mid- and long-term strategies and policies for food security and rural development.

In conclusion, China is in continuous transition to industrialisation and urbanisation, and is striving to maintain sufficient food supply. New measures are increasing the farmers' own productivity through developing the small farms, decreasing market regulations, and expanding international trade in food. Poverty reduction and social security programmes together with better education are parts of the efforts to strengthen agriculture and food security.

Current Schemes

China to Reform Grain Distribution System [5]

The first involves the opening of grain marketing areas. The government will allow grain prices in major grain marketing areas, including Zhejiang, Shanghai, Fujian, Guangdong, Hainan, Jiangsu, Beijing and Tianjin, to fluctuate in line with market supply and demand. This is aimed at restructuring the agriculture sector in those areas, and helping grain prices to rise in grain growing regions.

Second, protection of grain growing areas will be stepped up. Grain production and its economic returns will be ensured in major grain growing areas in China to protect farmers' interests. "This is the key to ensuring the country's food supply and giving full play to the comparative advantage in grain growing areas," Third, the responsibility system of provincial governments will be tightened. Local government officials should take full responsibility over local grain production and distribution. Fourth, the state will further enhance the functioning of its macro-control system. The central government will exercise blanket control over the grain distribution system, by protecting grain prices, establishing risk funds and improving the grain reserve system.

China will step up efforts in the construction of grain reserve warehouses as well as the information network, while increasing supervision of the market entry of grain marketing enterprises.

Sinograin [2]

China Grain Reserves Corporation (Sinograin) is a state-owned enterprise Founded in 2000 with a registered capital of RMB16.68 billion; Sinograin has grown into one of the largest and most wide ranging grain storage and transportation corporations in China and is a leader in the research and use of innovative technology and equipment. An old Chinese saying goes, "if you have sufficient grain in your barn, you can live your life without any panic." To guarantee the food security requires both the increase of production and the adjustment of grain reserves. That is the integration of ensuring grain production and keeping the reserves in the warehouses. The mission of China Grain Reserves Corporation (Sinograin) is to balance supply and demand for stabilising prices, protect farmers and safeguard the national food security.

Schemes Under Sinograin [2]

- Assume the task of governmental protective purchase of grain across the nation.

From 2005 to 2008, Sinograin had purchased a total of 320 billion catties grain. This resulted in income increase by more than RMB 8 billion for grain-growing farmers. The grains were to be used to build reserves for the country.

- Conduct the auctions to stabilise grain prices and increase market supply.

From 2005 to 2008, Sinograin sold a total of 220 billion catties grain and edible oil, to increase market supply and to control quick rise of grain prices.

- Undertake large-scale cross-regional deployment and balance of grain across the nation.

From 2005 to 2008, Sinograin transported 26 billion catties grain and edible oil to increase the grain storage in key distribution and grain-short areas.

- Assume national import and export task of grain and oil.

Sinograin assumes the tasks of acceptance and discharge of imported grain as well as the purchasing and exporting of some of domestic grain.

After nearly two decades of reform, the grain bureau still requires farmers to deliver specified quantities of grain. While it is too early to evaluate the impacts of 1998 new grain market reforms on grain production, price and marketing, issues related to the ability and willingness of local government to hold sufficient buffer stock for price stabilisation need to be addressed.

The Food for Work Policy [8]

Prior to the mid-1990s, the food-for-work policy was implemented in the form of payment in kind. At that time, the central government mainly provided commodities such as grains, cotton, clothing and low- and medium-grade industrial products as payment in kind for the building of roads and projects for drinking water supply for people and animals. Once a project was finalised, the local government provided funds to purchase construction materials, such as cement, explosives and reinforcing bars, and construction tools. Then it organised poor farmers to participate in the construction and paid them, as their wages, the commodities provided by the central government. At that time, the chief purpose was to provide relief to victims of natural adversities, and the food-for-work policy clearly had the purpose of relief, i.e., workers could directly receive income from their labour input. By contrast, improvement of rural infrastructure was of secondary importance; i.e., the objective of local development was a secondary consideration of the central government (Zhu Ling, 1995).

However, the expenditures deviated from the original intent as implementation proceeded. Most local governments had no money for the projects. They sold the commodities provided by the central government in order to purchase construction materials and to rent construction machinery for project construction. This practice substantially reduced the commodities that should have been distributed among farmers as their wages. In some areas, nothing was given at all. The element of "relief" in the policy was thus greatly weakened. The primary purpose of providing relief to the poor gave way to that of investment in rural infrastructure.

But such a change in policy created a dispute between the policy-making department and the local policy-implementing governments. The State Council's Office on Poverty Alleviation and Development insisted that "relief" should be an objective of the policy: The farmers should be paid for their labour input so that their households could benefit from participating in construction (Yang Zhong, 2003). However, local governments, due to their financial pressures, always sought to reduce wage payments to farmers, and tended to depend more on unpaid labour services. The practice of local governments of converting resources intended to pay for farmers' labour inputs into equivalent funds, does not appear to have been explicitly endorsed by the central government.

Under these circumstances, why did farmers continue to participate in construction activities for more than 10 years? What kind of reward did these poor farmers get for their labour? The answer is closely related to the basic system operating in rural China called contract system. First let us understand what it is.

The household contract system is practiced in rural areas of China, by which each person can contract a plot of land from the village for cultivation. The central government stipulated in a document issued in 1982 that the land-use right obtained by farmers should remain unchanged for the next 15 years. In 1996 when the first term of land contracts was over, the second term was given for 30 years. Moreover, it was explicitly stipulated that during the second term of contract, once the land was distributed, the determined ownership of land would not be altered for at least 30 years.

Legally speaking, the land belongs to the village, but farmers have the powers of management, disposition and inheritance of land. As long as farmers pay the contracting fee and agricultural tax, the villages cannot interfere in the farmers' operations on any account, let alone adjust or take back the cultivated land. The two provisions actually made the farmers' right to contracted management a long-term one and as a result, contracted land became an asset of farmers (Xiang Zhaolun, 2001).

Therefore answer to the above question is as follows. Capital construction on farms and water conservancy projects, implemented under the food-for-work policy, directly increases the asset value of cultivated land contracted by farmers. Again, the completion of roads and power grids in rural areas not only improves the living conditions of farmers, but also upgrades conditions of land management so that farmers can have more access to the market, which indirectly increases the value of contracted land. Although the farmers get no direct payment from their labour in implementing the projects, their land asset are considerably enhanced in value and farmers can secure better returns from managing the improved land resource in the long run. In this sense, the food-for-work policy has actually evolved into a policy by which the poor farmers provide labour which appreciates the value of their contracted land. The contracting of land, a practice unique in China, is closely related to the strong support of poor farmers for the food-for-work policy.

Thanks to the unique land system in rural China, 84.5 per cent of the peasant population owns land [9]. Only 15.5 per cent are landless. In India, the landless form 43 per cent of the population (according to India Rural Development Report, 1992, [10]). The farmers in China thus have a stable source of basic food supply which is not guaranteed to the landless labour of India. This in some way can explain the differences in malnutrition rates between the two nations. A Chinese farmer, so long as he is physically and intellectually capable, can manage to maintain the basic standards of living of his family by relying on a plot of contracted land.

Change in the Form of FFW Policy [8]

Since the mid-1990s, food shortages have been brought to control across the country. Through more than 10 years of efforts in poverty alleviation, food and clothing shortages has been greatly reduced among the overwhelming majority of the poor (Yang Zhong, 1997). Under these new economic circumstances, the central government could modify the food- for-work policy for poverty alleviation by allowing expenditures provided for in kind to be converted into money. Although the policy is still known as food-for-work, it has actually become investment to improve infrastructure in poor rural areas. Apart from roads and drinking water supply for human beings and animals, water conservancy works and power grids are also listed in the range of investment. Therefore, the policy should be defined more accurately as investment for updating the rural infrastructure in poor areas rather than the direct targeting at farmers.

Regarding the mode of investment, the government only pays for the technical work involved and for the machinery used in the construction. The labour input of farmers is basically unpaid. Farmers get their share, directly or indirectly, in the benefits generated from the completed projects. In other words, farmers do not get direct labour income from short-term employment in the construction of rural infrastructure; they can only increase their employment and incomes due to the development resulting from the improvement in the rural infrastructure.

It also needs to be mentioned that the farmers directly benefit only from participation in the building of water conservancy works. However, here too, there are differences in the quantum of benefits received among the villagers. Some benefit a lot due to the geographical proximity to the water conservation site while some, whose lands are at a distance, do not benefit much. In projects like the construction of rural roads and power grids, farmers differ greatly with respect to their ability to make use of the infrastructure. In the end, some farmers may not benefit at all (even though they participated in the construction), while some others would see substantial gains in their incomes.

Addressing Extreme Poverty and Child Malnutrition for 2011–2020 [4]

For the period of 2011–2020 the Chinese government initiated a new programme for rural poverty reduction, and at the end of 2011, it lifted the poverty line to 2,300 Yuan per capita annual net income equivalent to 1.8 USD per day at 2005 purchasing power parity. Measured with the new poverty threshold, the size of rural poverty is estimated at 128 million people, or 13.4 per cent of the total rural population.

In the past three decades, child malnutrition has almost been eliminated in cities. The nutrition monitoring team found out that the low-weight and stunting indices of the urban children under the age of five years had dropped from 5.3 per cent and 11.4 per cent to 1.4 per cent and 3.1 per cent during 1990–2005. Compared to children living in cities and non-poor rural areas, malnutrition of the children in the same age group but living in the poor rural areas remains a challenge. From 1998 to 2009, the low-weight rate of this age group declined from 14.5 per cent to 6.6 per cent, and stunting prevalence was reduced from 36.4 per cent to 18.3 per cent. However, in 2010, the stunting rate of the children aged from 12–24 months was still observed at a peak of 22.8 per cent.

China has adopted a multi- pronged approach to address the issues of food security, as can be seen from the following paragraphs.

- Maintaining food supply and demand balance through self-sufficiency in staple grains with imports of soybean and edible oil. In fact, China has already been importing about 40 million tons of soybean and around eight million tons of edible vegetable oil each year.
- Extending the social protection system from the urban to the rural. Since the end of 2009, under the experimental pension system, farmers aged over 60 years are eligible to receive a monthly non-contributory benefit of 55 Yuan from central government finance. Moreover, a rural minimum living standard guarantee system has been set up to target the extreme poor and it is financed by both the local and central governments. In 2012, 28 million rural households with more than 53 million people were assisted with a monthly average cash transfer of 104 Yuan per capita. The share of the transfer from central government accounted for 65.1 per cent and 60.1 per cent for the urban and rural schemes.
- Starting from spring semester in 2012, central government has undertaken a school feeding programme in 669 hard core poor counties that covers 26 million rural children. At the same time, an experimental programme of nutrition intervention is being implemented in 100 counties in ten provinces, with supplementary food for the rural children aged 6–24 months.

Lessons to Learn from China [3]

China, the world's most populous country, is highly recognised for its ability to feed over 20 per cent of the world population with only about seven per cent of the world's arable land. Despite extremely limited natural resources and a population that has doubled over the last four decades, per capita daily availability of food, the status of household food security, and nutrition have all improved significantly. Increased domestic production is almost solely responsible for increased per capita food availability. The growth of agricultural production since 1950, particularly after 1970, is one of the main accomplishments of China's development and national food security policies. China's experience demonstrates the importance of technology changes, institutional reform, incentives, rural development, and sector specific policies to ensure the adequacy of food supply through domestic production.

China's experience demonstrates that strong rural economic growth is a key element of absolute poverty reduction. Broad participation in rural economic growth reduced absolute poverty between 1978 and 1985. The country has successfully created jobs, both farming and non-farming (particularly in rural industry), for rural surplus labour since the early 1980s. Diversification of production and development of rural enterprises (TVEs) raised farmers' incomes and improved rural living standards.

Policies, including the grain reserve system, strong disaster relief programs, and large scale food-for-work schemes, have contributed substantially to the stability of China's food supply and access to food for the poor. China's experience also demonstrates the importance of relations between economic growth, political and social stability, and food security in a large country whose economy is undergoing fundamental adjustment and growing rapidly. Maintaining a relatively high level of food self-sufficiency is desirable in terms of both domestic and global food security.

Food self-sufficiency has been and will continue to be the central goal of China's agricultural policy. The Ninth Five Year Plan for 1996-

2000 and the National Long Term Economic Plan both call for continued agricultural production growth, rising rural income, continued high levels of food security, and the elimination of absolute poverty. The strategies to achieve the above targets include:

- Improving incentives for both public and private investments to enhance agricultural productivity (investments in land, irrigation, research, crop extension, etc.);
- Increasing farmers' income through the promotion of rural industrial development and off-farm employment;
- Strengthening anti-poverty programmes by raising poverty alleviation funds; and
- Emphasising economic as well as infrastructure development, and developing the state grain reserve system as a major government intervention to stabilise food supply and market prices.

China's research system has produced a steady flow of new crop varieties and other technologies since the 1950s. The robust growth of China's stock of research is in large part responsible for its dramatic agricultural growth rates.

Improved technology has been the largest factor, by far, behind grain production growth, and as such is a major source of increased food availability in China. Institutional reform and alteration of farmers' incentives have also brought significant agricultural production growth. Irrigation is another important determinant of China's agricultural growth in recent decades.

However, challenges remain. China is land-scarce and is short of water. Competition for land and water from non-agricultural uses, further environmental degradation, and declining efficiency as a result of poor land and water management can only worsen land and water supply problems. Policy makers have not yet addressed key issues including the unbalanced use of fertiliser, the inefficiency of the state monopoly in fertiliser market and distribution system, and fertiliser import policy.

The recent decline of the government's budgetary commitment to research has weakened the system. The agricultural technology extension system needs to be strengthened. Development of an efficient seed industry is not without roadblocks as appropriate supporting institutions (i.e., eliminating entry barriers, allowing the market to set prices, phasing out subsidies, and protecting intellectual property rights) take time to mature. The signs of poor land/water management and unclear land/water property rights have started to show in the form of structural deterioration and declining productivity. A better incentive system is needed to encourage both public and private investment in agriculture. Limited grain handling and internal transport capacity will be one of the predominating constraints shaping China's economy.

China's Food Security

From Self-Sufficiency to a Dual Strategy [6]

Ensuring basic food self-sufficiency has been the long-standing policy for China's food security strategy. Yet, as its domestic food production is increasingly falling short of the country's rising demand as a result of population growth, rapid economic development and urbanisation, self-sufficiency alone is no longer an option for China.

China is gearing up to the rising pressure to feed its increasingly affluent population. On the one hand, it still considers domestic food production the cornerstone of its food security strategy, and is determined to boost the country's domestic food production. On the other hand, China is turning to international sources to meet rising domestic food demand, especially in its quest to achieve cereal self-sufficiency.

Boosting Domestic Food Production

Investing in agricultural technology is considered a key to China's food security problems. It will increase investment and subsidies for the agricultural technology sector to improve land yield, resource efficiency and labour productivity. China has also been investing heavily in water and other rural infrastructure to increase its overall domestic food production.

Confronted with limited land and water resources, China is turning towards its "blue lands", including its territorial waters and EEZs, for food. At the recent 18th Party Congress Chinese leaders pledged that they would enhance China's capacity for exploiting marine resources. Chinese officials and marine experts advocate that the country's food system be more maritime-based. Development of aquaculture and offshore fishing is being prioritised.

Utilising Overseas Agricultural Resources

Domestic production alone, however, is insufficient to meet the country's rapidly rising demand, forcing China to gradually look outward for food. This outward-looking food security approach has three major aspects:

Firstly, Beijing will import more food from the international market. In the coming years, not only imports of soybeans, cotton, edible crops, sugar, dairy and other agricultural products will rise more cereal, particularly maize, will also be imported from the international market. To secure its food sources from overseas, and ensure stability in its food supply, China has been enhancing its agricultural trade ties with major food exporters, signing agricultural agreements with countries such as the United States, Canada, Thailand, Argentina and Australia.

Secondly, China has become a major player in purchasing or leasing land in foreign countries for agricultural production. Given the unreliability of the international food trade, China is increasingly concerned that over- reliance on the international grain market could undermine its food security. Hence, Beijing intends to boost the country's food supply by expanding its agricultural operations overseas.

The Chinese government has issued a raft of policies and regulations, offering support and guidance in terms of financial subsidies and concessions to Chinese enterprises investing in overseas agricultural projects. According to a study done by the grain industry in February 2012, China has acquired over 3.5 million hectares of land overseas in recent years.

Thirdly, as it relies more and more on the international market for food and as its overseas agricultural investment surges, Beijing is showing greater interest in reshaping international trade and investment rules. It realises that food self-sufficiency is not a practical and efficient approach to any country's food security. China's integration into the global food system is inevitable. In particular, China considers the deadlock of the Doha Round of WTO talks (2013) a main barrier to global agricultural trade liberalisation, which could potentially threaten the country's food trade and thus food security. Significantly, China is taking steps to push for a successful conclusion of the Doha Round.

Concluding Remark [3]

China has taken great steps to alleviate poverty since the late 1970s. Further poverty reduction has proven difficult because the remaining population in absolute poverty is widely dispersed in remote upland areas. Although nutrition has improved over the last two decades, malnutrition remains. In sum, China is generally capable of solving its food problems. The government may play a substantial role in improving both national and global food security. Full knowledge of present policies and their impacts on food production, stability of supply, trade, and household food security are essential to design appropriate policy measures and the institutional framework to improve food security. China's agricultural policies and their economic impact are of interest also to the rest of the world as the country is integrated into the world market.

References

1. Amanda Briney (2014),"The Geography and Modern History of China", in about education, geography.about.com. (http://geography.about.com/od/chinamaps/a/china.htm)
2. China Grain Reserves Corporation (2009), "Service Regulation" in About Us. (http://en.sinograin.com.cn/eng/shiming_Eng_sub2.jsp)
3. FAO Corporate Document Repository, Annex 3,"Agricultural Policy and Food Security in China"(http://www.fao.org/docrep/004/ab981e/ab981e0c.htm)

4. Ling Zhu (2013), "Agricultural policy reform and food security in China", Rural 21, 9.12.,2013 (http://www.rural21.com/english/a-closer-look-at/detail/article/agricultural-policy-reform-and-food-security-in-china-0000959/)

5. Xinhua (2001),"China to Further Reform Grain Distribution System" in china.org.cn, 22/8/2001. (http://www.china.org.cn/english/17892.htm)

6. Zhang Hongzhou (2013), China's food security : From Self -Sufficiency to a Dual Strategy -Analysis, in Eurasia review, march 19, 2013. (http://www.eurasiareview.com/19032013-chinas-food-security-from-self-sufficiency-to-a-dual-strategy-analysis/)

7. Zhang-Yue Zhou and Vasant P. Gandhi (2000), "Public Food Distribution System and Food Security of the Poor: The Indian and Chinese Experience Compared", Paper presented at the 44th Annual Conference of the Australian Agricultural and Resource Economics Society, the University of Sydney, Australia, 23-25 January, 2000(http://ageconsearch.umn.edu/bitstream/123746/2/Zhang%20Yue.pdf)

8. Zhu Ling and Jiang Zhongyi (2004),"The Food For Work Policy and Expansion of Rural Employment in Poor Areas in Western China", ILO, Geneva. (http://www.ilo.org/public/english/employment/recon/eiip/download/setp/setp16.pdf)

9. Qinglian, He (2011), China's over 100 million landless peasants", in the Epoch Times, February 4, 2011.http://www.theepochtimes.com/n2/opinion/chinas-100-million-landless-peasants-50605.html

10. Planning Commission (1992)), :India Rural Development Rport, 1992" http://planningcommission.nic.in/reports/genrep/arep9099/index.php?repts=ar91-92.htm

5

Vietnam

Compilation: Divy Thakkar

Introduction and History of Vietnam

Vietnam occupies the eastern and southern part of the Indo-Chinese peninsula in Southeast Asia, with the South China Sea along its entire coast. China is to the north and Laos and Cambodia are to the west. The government is a communist state [10]

The Vietnamese are descendants of nomadic Mongols from China and migrants from Indonesia. According to mythology, the first ruler of Vietnam was Hung Vuong, who founded the nation in 2879 B.C. [10]

Geographically, Vietnam is a small country compared to India. Vietnam has a land share 1/10th that of India. The proportion of population residing in urban areas is although comparable: 31.7 per cent in Vietnam and 31.16 per cent in India. [2]

Table 1: Population Figures for India and Vietnam

Vietnam	India	Parameter. (as of 2011)
87.94 Million	1.237 Billion	Population
68.3%	68.2%	Rural percentage
31.7%	31.8%	Urban percentage
330,957	32,88,000	Land Area (in sq km)

*[**Source:** The World Bank Report, World Bank Group,2013 (http://data.worldbank.org/country/vietnam)*

2. Economic and Demographic Indicators

GDP Growth Rates

According to the figures given in [15], Vietnam grew by about 5.25 Per cent per cent per year in last one decade. Our analysis of the Indian figures hasve shown that the Indian economy grew at a little faster rate at 7.18 per cent (calculated in the chapter on Bangladesh). However, if we see the GDP per capita figures, Vietnam's per capita income ($ 1911) is more than India's ($1499).

(http://data.worldbank.org/indicator/NY.GDP.PCAP.CD).

Since 1994, the per capita GDP of Vietnam has grown more than eight times compared to the fourfold growth of India. Two decades ago, Vietnam's per capita income was less than India's; but by 1990s, Vietnam surpassed India in its per capita income.

Table 2. Poverty Headcount Ratio of India and Vietnam.

Vietnam	India	Year
63.7	45.3	1990
21.4	37.2	2005
16.9	29.8	2010

*[**Source**: The World Bank Report, World Bank Group,2013 (http://data.worldbank.org/country/vietnam)]*

Similar story is repeated when we look at the poverty rates. 63.7 per cent of the population was living below the poverty line in Vietnam in 1990, as opposed to 45.3 per cent population in India. However, by 2010, the Indian poverty ratios declined by 34.2 per cent to become 29.8 per cent whereas the Vietnamese poverty ratios declined by 73.5 per cent to become 16.9 Per cent per cent. In two decades, Vietnam charged ahead from being a much poorer nation than India to a middle income nation, with almost half poverty rates compared to India.

Of course, the problem faced by India is much larger because the population of India is more than 14 times the population of Vietnam.

Table 3: Global Hunger Index of India and Vietnam

India	Vietnam	Year
32.6	25.7	1990
27.1	21.4	1996
24.8	15.5	2001
21.3	11.2	2011

*(**Source:** IFPRI, 2013 Global Hunger Index, 2013)*

Possibly because of its communist roots, Vietnam always had lesser hunger rates compared to India, as shown by Table 3. However, the fall in hunger rates also have been much faster in Vietnam: Vietnam has reduced its GHI by 56.4 per cent from 1990 to 2011 while India has a reduction of 34.6 per cent in GHI from 1990 to 2011. The Hunger index of Vietnam has been brought down due to the efforts of the government to ensure self-sufficiency in Agriculture. India's self sufficiency in food could not bring down its hunger rates because a large part of the rural population is landless that does not get to share in the gains of increases in agriculture productivity. Nor do the gains from increased service sector activities adequately reach the urban manual labour.

Table 4: Inflation Rate of India and Vietnam

2012	2011	2010	2009	2008	2007	2006	2005	2004	2003	2002	1998	1994	year
9.1%	18.7%	8.9%	7.1%	23.1%	8.3%	7.4%	8.3%	7.8%	6.9%	3.8%	8.8%	17%	Vietnam
7.2	8.2%	12%	10.9%	8.7	6.4%	6.1%	4.2%	3.8%	3.9%	4.4%	8%	10%	India

*[**Source:** The World Bank Report, World Bank Group, 2013 (http://data.worldbank.org/country/vietnam)*

India: Mean = 7.26 and Standard Deviation = 2.76

Vietnam: Mean = 10.37143 and Standard Deviation = 5.34695]

At the first glance, the inflation rates of Vietnam seem alarmingly high at times – 17 per cent in 1994, 23 per cent in 2008 and 18.7 per cent in 2011. The mean rate of inflation turns out to be in lower double digits, 10.37. The standard deviation is 5.3. India with its inflation rate at 7.26 and standard deviation at 2.76 has a much more stable economy compared to Vietnam. In spite of such wide fluctuations in prices, Vietnam has been able to manage its hunger situation better, as we saw in table 2 and 3.

Health Indicators

As far as health ratios are concerned, Vietnam is far ahead of India in IMR and MMR. India's IMR of 44 is worse by 2.5 times than Vietnam's 18 (Table 5) and its MMR of 220 is almost three times worse than Vietnam's 75 (Table 6). The life expectancy is about 15 per cent lower at 66 in India, while it is 76 in Vietnam (Table 7). As stated earlier, these ratios also, just like the poverty and hunger ratios might be better in Vietnam because of its earlier communist legacy. Communism affects the political freedom and economic activities negatively but seems to generate a positive impact on the provision of basic necessities.

Table 5: Infant Mortality Rates of India and Vietnam

Year	India	Vietnam
1994	80	31
1998	71	27
2003	60	22
2008	51	19
2012	44	18

[**Source:** *The World Bank Report, World Bank Group,2013 (http://data.worldbank.org/country/vietnam)*]

Table 6: Maternal Mortality Rates of India and Vietnam.

Year	India	Vietnam
2001	370	170
2006	280	162
2008	220	75

[**Source:** *The World Bank Report, World Bank Group,2013 (http://data.worldbank.org/country/vietnam)*]

Table 7: Life Expectancy at Birth of India and Vietnam.

Year	India	Vietnam
1994	60	22
1998	61	73
2003	63	74
2008	65	75
2013	66	76

[**Source:** *The World Bank Report, World Bank Group,2013 (http://data.worldbank.org/country/vietnam)] [9]*

Human Development Index

As all three components of HDI, (per capita income, literacy rates and life expectancy) are better for Vietnam, it comes as no surprise that the overall HDI index points hare higher for Vietnam compared to India.

Table 8: HDI Values.

2012	2011	2010	2005	2000	1995	1990	Year
0.617	0.614	0.611	0.573	0.534	0.48	0.439	Vietnam
0.554	0.551	0.547	0.507	0.463	0.438	0.410	India

Source: *Human Development Report 2013, The Rise of the South: Human Progress in a Diverse World. [5]*

Rice Exports

Vietnam has 33 million ha of land, of which rice occupies about four million ha. Rice is the main food of the Vietnamese people providing 80 per cent of the carbohydrates and 40 per cent of the protein intake in the diet. Because of the economic reforms and dissemination of technology, rice production has increased rapidly as can be seen from Table [12].

From Table 12 and 13 we can compare the production of Rice in both the countries. While Vietnam produces 30 per cent of what India produces, Vietnam exports 15 percent per cent of what it produces, India exports only around 1.5 per cent of what it produces. India enters the export market only now and then when it gets bumper crop; for example, in 2007, India exported more rice than Vietnam. Otherwise, it needs the entire production to satisfy its domestic demand. Vietnam produces much more rice than needed for its domestic demand. Rice is an important export commodity for Vietnam. With consistent trade policies and land regulation, Vietnam has been able to secure the place of the second largest exporter of rice in the world.

Table 9: Total Production of Rice for India and Vietnam.

2010	2009	2008	2007	Year
40,005,600	38,950,200	38,729,800	35,942,700	Vietnam
143,963,000	135,673,000	148,036,000	144.570,000	India

*[**Source:** Food Security Portal, IFPRI, 2013.*
(http://www.foodsecurityportal.org/vietnam)

Table 10: Export of Rice (in Million Tonnes)

Vietnam	India	Year
4,642,000	4,739,872	2006
4,558,000	6,450,062	2007
4,735,170	2,484,275	2008
5,968,762	2,148,040	2009

*[**Source:** Food Security Portal, IFPRI, 2013.*
(http://www.foodsecurityportal.org/vietnam)] [3] [4]

2. History of Rice Production

Mekong Delta is one of the most important regions growing rice. It once belonged to Cambodia till the late 17th century when it was taken over by Vietnam. The Cambodians still consider this area as "KamPucheaKrom or Lower Cambodia" not able to forget that they owned this area in the past. Cambodia even attacked Vietnam in 1978 to take control of the Mekong Delta. However, they were decisively defeated. This delta was the last part of the present day Vietnam that was annexed from Cambodia. [1]

Development of Mekong Delta is attributed to the French who colonised Vietnam in the middle of the 19th century with the basic objective of exporting rice grown in the delta to meet its large costs of colonisation. They developed a maze of canal system in the delta to grow three crops of rice in a year [1]. During the World War II when the Japanese occupied Vietnam and exploited the rich delta by exporting rice to their country, it denied nearly several million Vietnamese of their basic staple. Both France and Japan forcibly hoarded food from farmers to feed their troops. The inadequate food supply caused the famine in Vietnam; starting from 1943, peaking in March-May 1945 and continuing till the end of Pacific war (September, 1945). There was unprecedented starvation. Two million Vietnamese people were reported to have died of starvation. The famine was further compounded by unprecedented floods. In March 1945, Japan established the puppet Empire of Vietnam headed by TranTrang Kim. While this government tried to alleviate the suffering, they were unable to do so because Japan still stuck to its policy of hoarding food. [1]

At the end of the Vietnamese war in 1975, the country came under the Communist regime. Cooperative farming through communes was established. All rice growing fields of the delta became the state property the government also supplied to the farmers seed, fertiliser and other essentials. As this helped the poor farmers, it was popularly called "the iron rice bowl" – this allegory was meant to convey that "no one would get rich by this system, but a system where no one would fall through the cracks". Production was nationalised and marketing was strictly monitored: people wanting to buy and carry rice from the delta were not allowed to carry more than the ration of 80 kilograms (180 lb). [1]

Realising the ineffectiveness of the state control, the Government of Vietnam, in 1986, allowed the farmers "to grow and sell their rice". In 1994, the path breaking Land Law was passed. The government gave a lease of 50 years to the farmers to own the land and farmers were also allowed to sub-lease their land to others. This helped in transferring the lands from the cooperatives to the individual families. The farmers had then much better incentive to grow more rice. These developments resulted in an accelerated growth of the Mekong Delta and increase in income to the people. Consequently the delta is intensely populated now. The inhabitants of the delta are mostly constituted by the ethnic Vietnamese with Chinese, Khmers and Chams form the minority groups. [1]

3. History of Economic Policies

Vietnam was a country of food deficit during the seventies and eighties of the last century. As can be seen from Table [15], the agricultural production has been very low for the years 1968-1983. One reason can be the collectivisation of land. Since 1986, Vietnam started the "Renovation" policy, DoiMoi. It started transformation of Vietnamese economy from a centrally planned system into a market-oriented economy with a socialist orientation.

Table 11: Agricultural Production in Vietnam Over the Years (in Million Tonnes)

Production	Year
8,366,150	1968
11,125,000	1973
11,827,200	1976
11,362,900	1979
14,390,200	1983
19,225,104	1990
26,396,700	1996
32,529,500	2000
36,148,900	2004
38,729,800	2008
38,950,200	2009
42,398,346	2011

*[**Source:** FAO Statistical Wordbook 2013, FAO. (http://faostat.fao.org/http://faostat.fao.org/)]*

DoiMoi is generally recognised as a watershed in the country's economic and social development. Economic reforms accompanied by a shift from import substitution to an export orientation led to a period of significant growth. In 2004, after just under two decades, the GDP in constant price tripled, and the size of the economy reached about US$50 billion. The annual average GDP growth often exceeded seven per cent between 2001 and 2010. The growth was driven by the rapid expansion of the private sector (both domestic and foreign). By 2003, the private sector contributed about three-fifths of GDP and nine-tenths of employment. The trickle- down effect of this growth was also considerable: the proportion of population living under poverty line became 24 per cent in 2010. [13]

Comprehensive actions were made, including farm land reform, market economy transformation, investment in rural infrastructure (particularly irrigation schemes), promotion of agricultural research and provision of technology extension and credits to farmers. Thanks to these concentrated efforts, Vietnam not only became self sufficient in food but also became a leading food exporter: it is the second highest exporter of rice in the world, highest exporter of cashew nut and black pepper, and a major supplier of vegetables, fruits and fisheries products. [13]

Changes in demand have resulted in major structural changes in the agricultural supply sector, as output has shifted from bulk commodities, such as rice, to more profitable high-value products, such as fruit, vegetables, nuts, meat, and fisheries products. These changes were brought about in part by the economic reform process which stimulated the production of agriculture in three ways:

First, by expanding domestic incomes, the reforms have increased demand for fruit, vegetables, livestock, and fish, as consumers increasingly seek diversity and safety in their diets, particularly in urban areas.

Second, by establishing a realistic exchange rate and liberalising exports, the reforms have created new outlets for fruit, vegetables, and seafood processors.

Third, by raising the production and domestic availability of rice, the reforms have allowed rural households to allocate part of their land to fruits and vegetables with some assurance of being able to produce or buy the rice needed for household consumption.

4. Current Policies and Reforms to Ensure Food Security

Land Policies for agricultural land

The focus of the policies is firstly to ensure the proper use and management of paddy land. The government has set aside 3.8 million ha of land for growing paddy. This land is not allowed to be used for any other purpose. 3.8 million hectares of land would ensure paddy production of 41-43 million tons – this would meet the domestic demand and ensure the export volume of 4 million tons per year. In addition, there are also policies to protect and enlarge paddy land which include support policies for regions or provinces where favourable conditions for rice production and those for rice farmers exist. [12]

Limitations of this policy: The "rice first" policy of Vietnam might not be a very advantageous policy. If export is the aim, other cash crops might bring in more revenue. Also, environmental costs of monoculture, that too intensive rice cultivation that requires large amounts of water, can be very high.

Policies on Infrastructure for Agriculture

The policies concentrate on irrigation works as well as building storages. Irrigation has received the highest attention as it is the single most important way to increase the number of crops taken and for adopting the HYV technology. 75 per cent of total capital for agriculture in 1990s went into building irrigation facilities.

Rice exporters were encouraged to build warehouses and milling factories. Plant research and fertiliser production are also supported. The Government has been investing in building the new storage system and maintaining old storage system with total volume of four million tons. The capacity of new warehouse will reach the volume of 2.5 million tons. Government also allocates annually a budget of about two per cent GDP for science and technology. [12]

However, in general, the total investment in agriculture in Vietnam is still very small (about six-seven per cent of total investment in the whole country) and much lower than the contribution of the sector in national GDP. The implementation of project on building a four million tons of paddy storage system has not run as expected. Vietnam still needs to do a lot to support its agriculture.

Policies to Control the Price of Rice and Hence Ensure Food Security

To ensure fair returns to the farmers, government encourages companies to buy rice at a price that ensures at least 30 per cent profit to the farmers (This is in accordance to a bill passed in 2010). Based on the production cost survey of the Ministry of Agriculture and Rural Development, People's Committees of provinces and central cities determine rice price for each season and announce rice price in the provinces (paddy rice price equal = the production cost plus (+) 30 per cent profit) as a guide for enterprises to buy rice of farmers. When market rice prices fall below the price which is announced by the People's Committees of provinces and cities, Government establishes the rice price stabilisation fund and allows enterprises to use this fund to purchase rice. [12]

Future Plans of the Government

The Government plans to put an imposition on the enterprises to purchase about 2 Million Tonnes of rice for the three of months of August, September and October. The procurement of rice by these enterprises will help to build buffer stocks of rice (for releasing the rice in the market if the price rises or so as to be an aid during Emergencies). The procurement of rice ensures that the price of rice in the market is at par with the price of Provincial People's Committees. Government will give 100 per cent of interest of loans for the enterprises to enable them to purchase rice for temporary storage for the three months mentioned above. [12]

Policies on Rice Export

Restricting the Export in Emergency Case

A strict policy to set export limits was implemented in 2008 because of the global economic crisis, the objectives of this policy are to constrain inflation, stabilise domestic price consumption, and ensure food security of the local people before exporting rice to importers such as Philippines. This policy considerably affects the rice export enterprises because changes in their export plans depend on export management scheme in each period, so it's difficult to make long-term plans. Also, the sudden contract termination causes to reduce prestige of Vietnam enterprises with their partners. At the same time, many enterprises only focus on Government to Government contracts; and there is no motivation to expand market.[12]

During the global rice crisis in 2008, the government announced that no export contracts would be approved unless the exporter already had 50 per cent of the sale in stock. Other requirements were imposed: prices had to be in line with the export tax; and shipment had to occur within 60 days of signing. Additionally, export targets were revised for 2008, with a January to June quota of 2.25 million tons: 50 per cent of total rice exports for 2006 and 2007 combined. Even with new restrictions in place, domestic prices still remained volatile. [7]

Regulating Conditions for Rice Export Enterprises

Because of unfair trading in the market, the government decided to allow only a few enterprises to be the players in the market. Conditions for storage capacity, stocks of reserves and milling facilities were laid down for selecting the enterprises. This was done to control the exports When the domestic rice price increases suddenly, the ministries can guide traders to sell the reservation rice into the domestic market. [12]

Limitations of this Policy

The policy does not reach the macro goal because the most important issue is innovation of production method aiming to link "four stakeholders" (enterprise – farmer – scientist – government) with farmers and enterprises as the main forces.

There is a danger of centralisation of power also when entry into a market is restricted. The enterprises would need to have large and long term contracts acquired. They will also have to invest in material zones that would not only procure but sell quality inputs and help in extension of technologies. The conditions of the Decree will only help the existing big companies to increase their power and eliminate the small businesses because the small businesses cannot meet the requirements of storage and milling capacity. The Decree will also be a barrier to the new companies which want to go into the business of rice sector. [12]

Direct Support to the Paddy Farmers

Support Annually for the Farmers

- Cash support of 500,000 VND/hectare/year for organisations, households and individuals which grow rice on the rice specialised land. [12]
- Cash support of 100,000 VND/hectare/year for organisations, households and individuals produce on other lands except upland rice which was expanded land out of planning. [12]

Support for paddy farmers damaged by natural, epidemic diseases

- Support 70 per cent of cost of fertilisers and pesticides for rice production for farmers when they lose over 70 per cent of production. [12]
- Support 50 per cent of cost of fertilisers and pesticide for rice production for farmers when they lose over 30-70 per cent. [12]

Support for reclamation and improvement of rice land

- Support 70 per cent of reclamation and improvement costs for farmers when they reclaim unused land into rice land or improve other land into wet rice specialised land; the cost will be determined by provincial People's Committees. [12]
- Support 100 per cent of rice variety cost in the first year which is grown on new reclamation land. [12]

Supporting frameworks

- The localities which receive from the central budget will be supported 100 per cent of fund. [12]
- The localities which receive less than 50 per cent from revenues of the central budget will be supported 50 per cent of fund. [12]

Limitations of This Policy

(i) The difficulty to implement and monitor because a large number of farmer households, approximately 10 million households spread through the eight regions and 63 provinces of whole country

(ii) Difficulty to identify whether upland rice area is in planning area or not

(iii) The inefficient impacts do not make a considerable change in income of paddy households compared to other farmer households, or households assigned to other farmer households, or households assigned to work in non-agricultural sectors. [12]

Direct Food Support

Poor households involved in afforestation and forest protection will receive 15 kg rice per capita per month during the period when they are not able to provide themselves with staple food (not over seven years). Poor household in the border areas will be granted 15 kg rice/person/month until they become self- sufficient in food. Direct food support for provinces that suffered from natural disasters. [12]

5. The Debate on Landlessness

Freeing up agricultural land markets is a risky reform for the transition economies of East Asia, where most of the poor rely on agriculture for their livelihoods. While such a reform is expected to promote economic efficiency, policy makers have worried that it would undo socialism by recreating a rural proletariat-a class of poor rural workers. This concern has clearly inhibited liberalising agricultural land markets in China, despite the likely efficiency gains. However, other East Asian transition economies, including Vietnam, have nonetheless embarked on this reform and established de facto private ownership of agricultural land. [6]

Starting in the late 1980s, agriculture in Vietnam was de-collectivised and the land was allocated across households by administrative means and in a relatively equitable way (Ravallion and van de Walle, 2004). Since then, legal reforms have supported the emergence of a land market. The 1993 Land Law introduced official land titles and permitted land transactions for the first time since communist rule began. Land remained the property of the state, but usage rights could be legally transferred and exchanged, mortgaged and inherited. Since these reforms, there have been signs of sharply rising rural landlessness, which have fuelled much debate about the wisdom of the reforms [6].

The debate is centered around on if the country is heading towards a South Asian style of rural development in which there is a large and unusually poor landless class? Or are farmers simply selling their land to pursue more rewarding activities? In short, does rising rural landlessness in the wake of market-oriented reforms signal an emerging new poverty concern for Vietnam, or is it simply a by-product of the process of poverty reduction? The Hanoi-based Center for Rural Progress (2004) argued that an active land market in the Mekong Delta contributed to more rapid poverty reduction by allowing more efficient farmers to accumulate more land, fostering diversification and increasing farmers' access to credit.

There is evidence that land allocation has become more efficient since the 1993 land law. Deininger and Jin (2003) found evidence that more able farmers acquired more land after the reform. Ravallion and van de Walle (2005) found that agricultural land was re-allocated in a way that attenuated the initial inefficiencies in the administrative assignment of land at the time of de-collectivisation; households that started with an inefficiently low (high) amount of cropland under the administrative assignment tended to increase (decrease) their holdings over time. [6] Ravallion and van de Walle (2005) also found evidence of polarisation among those who started off with too little land; while most of these households acquired more land, a minority sold or transferred all their farmland, possibly to take up other non-farm activities or to pay off debts. Landlessness stemmed in part from inefficiencies in the initial administrative allocation. [6]

Critics of land-markets have been concerned that the poorest become landless and hence dependent on wage labour, which (it is believed) makes them worse off. This view stems in part from a mistrust of labour markets. For example, in coming to the conclusion summarised above, Dong alludes to an efficiency-wage argument (citing Dasgupta and Ray, 1986, and Moene, 1992), whereby workers with too few assets end up unemployed (although Dong does not establish that the poor would be better off without land markets). [11]

Table 16. Landlessness vis-a-vis Poverty

Poverty % of Landless population.	Landless Population %	Poverty % of population with land	Population with Land %	Year
50.87%	7.8%	70.04%	92.2%	1993
40.51%	6.9%	45.90%	93.1%	1998
25.11%	13.9%	38.60%	86.1%	2002
18.14%	12.3%	25.99%	87.7%	2004

[Source: Martin Ravallion and Dominique van de Walle: Does Rising Landlessness Signal Success or Failure for Vietnam's Agrarian Transition? Development Research Group, World Bank. (2006)]

The unemployment rate among the landless is low. Based on adults who report that they did not work because they could not find work in the last 12 months, only 2.6 per cent of landless rural households in 2004 had an unemployed adult, and this was almost identical in 1993 (2.5 per cent). [14]

The table above provides summary statistics. For rural Vietnam, the landlessness rate increased by two-thirds over the period, to slightly more than 12 per cent in 2004. (These are population weighted; the proportion of households who were landless rose from 8.4 per cent to 13.6 per cent.) The rise in landlessness naturally put upward pressure on the inequality of land holdings. In all years, the poverty rate is higher for those with land than for the landless. Similarly, mean consumption is higher for the landless. There was a sharp contraction in the incidence of poverty, which occurred at roughly the same rate for those with land as the landless. Over 1993-2004, the proportionate gain in consumption was higher for those with land; the ratio of the mean consumption for the landless relative to those with land fell between 1993 and 2004 from 1.33 to 1.24, though within the period it fell (to 1.21 in 1998) then rose (back to 1.33 in 2002). [6]

There have been two very different interpretations of rising landlessness in the wake of Vietnam's land market reforms. The first says that, starting from a relatively equitable allocation of land, a rural class structure emerged as rich farmers bought land from poor farmers, who then became the poor landless labourers found at the bottom of the new class structure. By this view, the main dynamic in the transition process that has led to rising landlessness is a structural shift in the relationship between landlessness and living standards, whereby, in the wake of the reforms, the probability of being landless rises among the poor. [6]

The second interpretation assumes that rising living standards in a developing economy inevitably entail a partial shift out of farming. With sufficiently high returns to schooling it is more likely to be the non-poor who tend to be landless. By this view, there need not be any change over time in the relationship between the probability of being landless and one's standard of living. Rather, both rising landlessness and falling poverty happen in tandem, as some farmers choose to sell their land to take up new opportunities. Instead of a "structural" shift in the relationship between landlessness and poverty, rising landlessness and falling poverty jointly reflect a process of economic transition made possible by the introduction of land markets [6].

With the means of findings stated above, there is no conclusive proof for a transitive relationship between the rising landlessness and poverty in Vietnam. However, the critics of the land reform still hold the privatisation of land as one of the key contributors to poverty. "In reality, studies point towards the existence of a trend where people are moving towards paid stable jobs than being employed in Agriculture."

References

1. Clark D Neher (2010). Vietnam War at Crossroads: An Introduction to South Asia. Center for Southeast Asian Studies, Northern Illinois University. April 2010. (http://www.seasite.niu.edu/crossroads/cneher/cn.vietnamwar.htm)
2. Dr. C. Chandramauli (2011). Rural Urban Distribution of Population. Census of India 2011. http://censusindia.gov.in/2011-prov-results/paper2/data_files/india/Rural_Urban_2011.pdf
3. Food Security Portal, IFPRI, 2013. (http://www.foodsecurityportal.org/vietnam)
4. Food Security Portal, IFPRI, 2013. (http://www.foodsecurityportal.org/india)
5. Human Development Report 2013. The Rise of the South: Human Progress in a Diverse World. UNDP.
6. Martin Ravallion and Dominique van de Walle (2006). Does Rising Landlessness Signal Success or Failure for Vietnam's Agrarian Transition? Development Research Group, World Bank. (2006) (http://elibrary.worldbank.org/doi/pdf/10.1596/1813-9450-3871)
7. Murray Folton, Travis Reynolds (2012). The Political Economy of Food Price Volatility: The Case of Vietnam and Rice. University of Sakatchewan. (http://www.spaanetwork.fsaa.ulaval.ca/uploads/tx_centrerecherche/SPAA_PB_Political_Economy_of_Food_Price_Volatility_-_The_Case_of_Vietnam_and_Rice_01.pdf)
8. Prime Minster of Vietnam (2009). Speech during The World Summit On Food Security. November, 2009. (http://www.fao.org/fileadmin/templates/wsfs/Summit/Statements_PDF/Tuesday_17_PM/33_Viet_Nam_171109_PM.pdf)
9. The World Bank Report (2013), World Bank Group. (http://data.worldbank.org/country/vietnam) http://data.worldbank.org/indicator/NY.GDP.PCAP.CD
10. The World Factbook 2013. Central Intelligence Agency, (https://

www.cia.gov/library/publications/the-world-factbook/geos/vm.html)

11. Thomas Sikor. Land Allocations in the Vietnamese Uplands: Negotiating Property and Authority. Digital Library. Indiana University.

    (http://dlc.dlib.indiana.edu/dlc/bitstream/handle/10535/1225/Sikor_228001.pdf?sequence=1)

12. Tran Cong Thang (2014): Food Security Policies Vietnam. Asia Pacific Information Platform on Agricultural Policy. Food and Fertiliser Technology Center for Asia and Pacific Region. 14.03.2014. (http://ap.fftc.agnet.org/ap_db.php?id=212)

13. Vietnam Food Safety and Agricultural Health Plan, 2006. (http://siteresources.worldbank.org/INTVIETNAM/Resources/vietnam_sps_report_final_feb_06.pdf)

14. Viet Nguyen and Tim McGrath and White Pamela (2006): Agricultural Land Distribution in Vietnam: Emerging Issues and Policy Implications. MPRA Paper No. 25587. Munich University Library, Germany. (http://mpra.ub.uni-muenchen.de/25587/1/MPRA_paper_25587.pdf)

15. http://www.tradingeconomics.com/vietnam/indicators

6

Thailand

Compilation: Rohan Sharma

Introduction

Thailand, officially the Kingdom of Thailand formerly known as as Siam is a country at the centre of the Indo-China peninsula in Southeast Asia. With a total area of approximately 513,000 km2 (198,000 sq. mi), it is the 20th most populous country in the world, with around 65 million people. Thailand is an export dependent economy with exports accounting to two-thirds of its GDP. Agriculture accounted to 8.4 per cent of the entire GDP in the year 2012 [24, 2]

Thailand ranked second in the best quality of life in ASEAN. It became an upper middle income nation in 2011 notwithstanding political uncertainty and volatility [27, 19]

Over the past 50 years, Thailand has developed from a low-income nation that produced only basic commodities to a middle-level, diversified economy that has balanced strengths in agriculture, industry and services. In 2012, according to the Office of the National Economic and Social Development Board, Thailand had a GDP of THB11.375 trillion (USD366 billion). Thailand's high economic growth at eight-nine per cent per year during the late 1980s and early 1990s was interrupted by the "Asian Crisis" of 1997-1998; robust growth at around 5 per cent from 2002 to 2007 was again slowed down by the global financial crisis of 2008–2009.[13,27,22]

Indicators of food Security

Economic Indicators

Between 1980 and 2011, the Thai GDP grew from 913.7 billion Thai bahts to 4600 billion baths. Indian economy is much larger than the Thai: it is almost 10 times larger. The Indian economy also grew well during the same period from 8561.63 Thai Bahts to 55,558.31 Thai Bahts. The average rate of growth from 2005 to 2011 for India has been 8.39 and it was 3.26 for Thailand. The standard deviation was slightly lower for India at 2.19 and slightly higher for Thailand at 3.26. The paired t test shows that India's growth rates for these years were statistically significantly higher than Thailand–with t statistics at robust 7.919 and p value at 0.0002. Thus of late, Indian economy has been growing faster than the Thai economy, although the overall standard of living is much higher for Thailand, as mentioned in the introductory paragraphs.

Table 1: GDP Growth Rate

Year	1980	1985	1990	1995	2000	2005	2006	2007	2008	2009	2010	2011
Thailand	4.60	4.64	11.62	9.24	4.75	4.64	5.09	5.04	2.48	2.33	7.810.	077
India	3.63	4.89	5.63	7.35	5.71	9.05	9.39	10.08	6.19	5.04	11.23	7.75

*2008 was the year of financial crisis for the world. In 2011, Thailand was hit by floods. After 1999 Asian crisis, the Thai economy never went back to the old rates of growth.

***(Source:** IMF's World Economic Outlook Database of April 2013). [6]*

Thailand: Average Growth rate: 3.258

Standard deviation in growth rates: 3.443
India: Average growth rate: 8.39
Standard deviation in growth rates: 2.192

Food Security and Health Indicators

The eight Millennium Development Goals (MDGs)–which range from halving extreme poverty rates to halting the spread of HIV/AIDS and providing universal primary education, all by the target date of 2015–form a blueprint agreed to by all the world's countries and all the world's leading development institutions. The deadline for each nation to achieve the MDGs is 2015. Thailand is on track to meet all or most goals including eradication of extreme poverty, providing universal primary education, reducing child mortality and improving maternal health, promoting gender equality and responding to HIV and AIDS, malaria and other diseases. [16]

The maternal mortality and under-five mortality rates have been greatly reduced and more than 97 per cent of the population, both in the urban and rural areas, now have access to clean water and sanitation [27].

If we compare India to Thailand, the indicators show that Thailand is far ahead of India in all the indicators of food security. The malnutrition among the children below age five is one third in Thailand (seven per cent compared to 21.3 per cent in India), IMR is one fourth (11, as opposed to 44 in India), MMR is almost 700 per cent more in India (190 as opposed to just 26 in Thailand).

The figures are equally in favour of Thailand for literacy (93.5 per cent as opposed to 74 per cent in India), poverty rates (13.2 compared to 21.9 in India) and Global Hunger Index scores (8.1, compared to 21.3 in India).

In addition to its economic status, the high figures reflect the investment that Thailand seems to have made in its social sectors. Although the public spending in education is similar for the nations, (3.8 per cent and 3.2 per cent for Thailand and India respectively), the public health expenditure is double that of India.

Table 2: Comparison of Food Security Indicators Thailand-India

	Thailand (1990)	India (1990)	Thailand (2013)	India (2013)
Malnutrition in children below age five (weight/age)	17.4	NA	7	21.3
Infant mortality rate (per 1000 live births)	31	88	11	44
Maternal mortality rate Poverty Head count	NA	NA	26	190
(below national poverty line	58.1	NA	13.2	21.9
Global hunger index	15.1	32.6	8.1	21.3
Literacy rate	NA	52.21	93.5	74
Unemployment rate	2.7	4.1	0.7	3.4
Public spending on education (Per cent of GDP)	NA	NA	3.8(2010)	3.2(2010)
Public health expenditure	NA	NA	76.4*	33.1*
Inflation rate	5.9	9	2.2	10.9
Life expectancy at birth	59	70	74	66

* The values mentioned are the percentages of money invested by government out of the total amount spend on healthcare. For example, if the total expenditure on Healthcare in India is 100 million dollars, 33.1 million dollars are spend by government and the rest is private funded

(Source - http://data.worldbank.org/indicator)

Poverty in Thailand

Poverty in Thailand is primarily a rural phenomenon, with 88 per cent of the country's 5.4 million poor living in rural areas. Some regions-particularly the North and Northeast and some ethnic groups lag greatly behind others, and the benefits of economic success have not been shared equally, especially between Bangkok, Thailand's largest urban area, and the rest of the country.[27]

From the 1970s until the early 1990s the poor were mostly agricultural workers in the rural areas, particularly in the north, northeast, or southern regions, whose highest level of education was six years of primary schooling [23].

Investment, goods and services, government programmes, and well-paying jobs are still largely concentrated in the capital and nearby urban areas despite government efforts to decentralise economic development. The concentration of resources and investment in Bangkok is evidenced by the fact that it contributes over 50 per cent of the country's GDP despite hosting only about 10 per cent of the population [23].

Agriculture and Agricultural Exports in Thailand

Thailand was the second largest exporter of rice till 2011. Other agricultural commodities produced by Thailand in significant amounts include fish and fishery products, tapioca, rubber, grain, and sugar. Exports of industrially processed foods such as canned tuna, pineapples, and frozen shrimp are on the rise. India produces more rice than Thailand but has a much larger population to cater to; therefore India is not a steady exporter of food grains in the world market.

Rice production increased significantly in the past two decades in Thailand (from 17 million tonnes to 37.9 million tonnes. The production has been relatively stagnant in India during the same period. Thailand has been using the modern methods of cultivation whereas most parts of India grow rice the traditional way.

Table 3: Agricultural Growth of India and Thailand

	India (1992)	Thailand(1990)	India (2011)	Thailand (2011)
Rice production	111 m tonnes	17 m tonnes	104.32million tonnes	37.9 m tonnes
Rice exports	0.9 m tonnes	4 m tonnes	7.3 m tonnes	10.7 m tonnes
Agriculture contribution to GDP	29	12	18	13

(**Sources** - *http://data.worldbank.org/indicator*
http://en.wikipedia.org/wiki/Rice_production_in_Thailand
http://en.wikipedia.org/wiki/Rice_production_in_India
http://www.indiastat.com/agriculture/2/foodgrains/17180 rice/17194/stats.aspx
http://oryza.com/content/india-fy-2012-13-rice-export-earnings-stand-56-billion-40-previous-year
http://www.bloomberg.com/news/2012-12-17/rice-production-in-thailand- poised-to-reach-record-in- 2012-2013.html)

Historical Perspective

Thailand is the only Southeast Asian nation that has never been colonised. Prior to 1932, the Kingdom of Siam did not possess a legislature, as all legislative powers were vested within the person of the monarch. The history of Thailand from 1932 to 1973 was dominated by military dictatorships which were in power for much of the period. For most of the 1980s, Thailand was ruled by Prime Minister Prem Tinsulanonda, a democratically-inclined strongman who restored parliamentary politics. First constitution came into Thailand in 1997 and the 2001 elections in Thailand were the first general elections under the 1997 constitution [5].

The Thai economy collapsed as a result of the 1997 Asian financial crisis. Within a few months, the value of the baht floated from THB25/USD (its lowest point) to THB56/USD. The Stock Exchange of Thailand (SET) dropped from a peak of 1,753.73 in 1994 to a low of 207.31 in 1998 [13].

The country's GDP dropped from THB3.115 trillion at the end of 1996 to THB2.749 trillion at the end of 1998. In dollar terms, it took Thailand as long as 10 years to regain its 1996 GDP. The 2006 Thai coup d'état took place on Tuesday 19, September 2006, when the Royal Thai Army staged a coup d'état against the elected caretaker government of Prime Minister Thaksin Shinawatra. The coup d'état, which was Thailand's first non-constitutional change of government in fifteen years, followed a year-long political crisis.

From virtual collapse during the Asian crisis of 1997-98, Thailand has managed to weather political storms, including the disruption caused by the military coup of 2006, and a disastrous and calamitous tsunami, to retain its status as an emerging Asian giant. [20]

History of Food Security Policies in Thailand

1. Surveys (Early Period)

Dietary surveys were conducted as early as 1931 with the assistance of foreign nutritional experts. Zimmerman found that although there were no serious dietary deficiencies, diets were mostly rice and fish with very little meat, dairy products and fat. In 1934–1935, the Siam Second Rural Economic Survey was carried out by Andrews. He found that the impact of the 1932 economic depression on food expenditure was obvious but there was little evidence of actual dietary deficiency except the lack of variety in the North and the North–East. [8]

2. Pre-National Food and Nutrition Plans Period (1961-1976)

1. Between the years 1960 and 1970, development programmes (National Economic and Social Development Plan (NESDP) in Thailand led to a rapid annual rate of growth in the GNP, averaging 8.4 per cent
2. The First, Second and Third NESDPs (1961–1976) followed the health approach stressing remedial actions in the medical institutions.[21]

3. Developmental Processes

1. There was a carefully planned national nutrition survey conducted in 1960 by a team of Thai and United States experts.

Some of the major findings in this survey were as follows:

- Protein-energy malnutrition was the most severe problem, especially among pregnant and lactating women, infants, and preschool children in poor urban or rural areas.
- Vitamin A, thiamine, and riboflavin deficiencies and iodine-deficiency goitre were found in the northern and north-eastern regions.
- Anaemia was also high, but the aetiology was not clear. Dietary intake of iron was high (but the issue of bio-availability was not known at the time).
- A high prevalence of urinary bladder stone disease, a disease whose cause was then unknown, was found in the northern and north-eastern regions.

Among the specific recommendations included in the report was the better co-ordination of (i) the different activities related to nutrition such as those of the Nutrition Division of the Ministry of

Public Health; (ii) the teaching and research activities of the School of Public Health and the medical schools; (iii) the food and beverage laboratories of the Ministry of Industries; and (iv) the development activities of the Processed Food Organisation and the Army Quartermaster Subsistence Division.

2. There was the *Expanded Nutrition Project,* a pilot project initiated in March 1961 in 10 villages in the north-eastern province of Ubon Ratchathani. The key aims were to encourage and mobilise rural communities to grow and use local food more to improve their diets, and the co-operation and co-ordination between various disciplines including health, agriculture, education and community development, [10]

During this preparatory phase before the First National Food and Nutrition Plan, nutrition problems and target groups were identified and prioritised, pilot projects were tested, a critical mass in terms of manpower in various ministries was formed, and tremendous international support obtained.

4. Nutrition-Related Policy and Health Development

1. Poverty Alleviation Plan (PAP).
2. Basic Minimum Needs and Quality of Life Movement.
3. Growth Monitoring, and Food and Nutrition Surveillance System (FNSS).

Poverty Alleviation Plan (PAP)

Four key programmes were implemented:

(i) *Rural job creation programme:* Jobs were created for rural people during the dry season to boost their income. Most of the employment was given to people in the locale so that rural people would remain in their communities and participate in community development activities.

(ii) *Village development projects or activities:* The activities included village fish ponds, water sources, prevention of epidemic diseases affecting poultry, cattle and buffalo bank, and other development projects focused on rural poor to improve their economic status and household food security.

(iii) *Provision of basic services:* Public services for rural poor such as health facilities and health services, nutrition, clean water supplies, and literacy education programmes were directed to the targeted areas.

(iv) *Agricultural production programme:* Important programmes included nutritious food production (especially crops used for supplementary feeding of young children), upland rice improvement projects and soil improvement projects.[11, 12]

Basic Minimum Needs and Quality of Life Movement

A set of Basic Minimum Needs (BMN) and their indices were developed in 1981. They were to be used by the villagers themselves.

The results from the process are used to formulate a village proposal and budget allocations are made from the centre based on such proposals.

Growth Monitoring in Primary School and Food and Nutrition Surveillance System

Growth monitoring started in the First National Food And Nutrition Plan (1977–1981) as a means to watch over and identify high-risk children at the earliest possible stage. During the second National Food And Nutrition Plan (1982–1986), nationwide Primary Health Care offered the opportunity for widespread *growth monitoring activities*.

Several evaluation studies at the beginning of the third National Food And Nutrition Plan (1987-1991) indicated weaknesses at various steps of the growth monitoring process. Important weaknesses were: the incomplete census of children under five in most communities made it impossible to calculate the percentage of children the programme covered; VHV (village health volunteers) tended to focus on getting the weighing done quickly to shorten the mothers' waiting time thus seldom interpreted results to mothers or caretakers; and worst of all, mothers of children with moderate to severe malnutrition received no advice. [18, 12, 26]

The FNSS information system has sought to cover data on directly and indirectly related factors: it covers data on weather conditions (amount of rainfall, and soil fertility); agricultural production (crops, livestock and fishery); economic indicators (household income, wage rate); local consumer price index, health conditions (respiratory tract infections, acute diarrhoea and dengue haemorrhagic incidence) and nutritional status (birth weight, pre-school and school children). Recently, two additional indicators have been added: the non-agricultural income of the household and food consumption as reported by school children on recall to teachers from consumption the previous day. Such monitoring would help in identifying any kind of short fall or danger signal.

Other Important Nutrition Interventions

Supplementary Food Programmes

To reduce the dependence on manufactured nutrition supplements, during the fifth NESDP, the strategy changed to village-based supplementary food processing. Supplementary food formulas of rice, beans and groundnut or sesame were developed. Each formula of 100 g of the supplementary food provided approximately 450 Kcal and 12-14 g protein. These food mixtures could be processed in the village using simple, low-cost equipment and operated by trained villagers.[9]

A new strategy of *food coupons* was then introduced in addition to the village food processing in 1988. The food coupon was given to individual children who were second and third degree malnourished. A monthly booklet of thirty coupons, each worth 3 baht in 1988 was given to the mothers of these children (5 baht in 1991). Every day, one coupon could be used at the local shop on specific items of food indicated on the coupon, such as eggs. Authorised shop owners in the village collected the coupons and were reimbursed from the sub-district health offices.

School Lunch Programme

National and government expenditure on education and lunch programme increased in real terms during the decade. Since the economy expanded rapidly at the end of the 1980s, the share of education in GDP has slightly dropped, although it has still increased in real terms.

(a) Bangkok Metropolitan

The school lunch programme of the BMA was administered by the school committee. Teachers were responsible for the menu and preparation of the foods. Either simple and nutritious Thai meals (such as rice and Thai chicken curry, chicken noodle soup, etc.) or a snack at mid-morning or afternoon breaks (such as soybean milk, boiled mung beans with sugar added, etc.) were provided. The menu was planned with the nutritional content, and the use of low price protein sources, such as soy beans in mind.

Present day School Lunch

School lunch covers all public kindergarten and elementary schools, including those in remote rural areas (approximately 30,000 schools). It has benefitted around two million primary school students and 700,000 pre-schoolers as of 2011. [15]

SLP runs for 200 days of the school year. As of 2004, the programme has allotted 10 Baht/child, multiplied by 200 days, the total annual budget per child is 2,000 Baht. In the beginning, SLP focused on the quantity of meals produced; once that was achieved, the focus shifted to the meal quality and nutritional content. [15]

Summary

The target group of this scheme was very elderly Thai people (60 years of age or older) who are not served by elderly public facilities or do not currently receive income (i.e., government pension recipients, government employed persons). There were about 6.87 million (approximately 95 per cent of the elderly) such people.

The number of registered elderly was 5.65 million (82.2 per cent of target population) who were given in-cash benefits of 500 baht ($ 31.4 at 2005 price level) per month. In principle, the elderly or the authorised representative must register with the local authorities where he or she has inhabitancy registration.

The qualified recipients can choose among four methods:

- To receive cash directly from the local authority office.
- To designate an authorised representative to receive cash directly from the local authority office.
- To have the pension transferred to the elderly person's bank account.
- To have the pension transferred to a bank account of an authorised representative.

The total expenditure was around 33,917 million baht in fiscal year 2010 or 2,129.1 at 2005 prices. That translated into 0.37 per cent of GDP which was funded by General tax revenue.

Current Policies

1. 500 Baht Universal Pension Scheme

Growth Story of the Universal Pension Scheme

The old-age allowance system was established in 1993. In the first year after the introduction of this scheme, the number of recipients was merely 20,000. In 1995, recipients had increased to 110,850. In 2000, the allowance amount was 300 baht per person per month. In 2005, further to Thailand's decentralisation process, the tasks of defining clients and defining allowance payments were delegated to local authorities through grants from the central government. In 2006, benefits were increased to 500 baht per person per month. However more than 50 per cent of underprivileged elderly did not receive the old-age allowance, according to the Monitoring and Evaluation Project of the National Elderly Plan by the College of Population Studies, Chulalongkorn University. Implementation failure forced the Government to go from a means- tested to a universal scheme which officially started in April, 2009. As of fiscal year 2010, recipients of the 500 baht pension represent approximately 77.5 per cent of the elderly population. There are still 1.22 million elderly who have not yet registered for the Scheme.[17]

2. Rice Pledging Scheme

The Thai government, led by Prime Minister Yingluck Shinawatra, implemented the rice pledging scheme immediately after the 2011 election, to enhance the support to Thai farmers. The government for some reason decided to go overboard and' started paying paddy farmers 15,000 baht (US$420) per ton–a 60 per cent increase over 2010.[7]

To give such high prices to farmers, the rice subsidy policy, known domestically as the rice-pledging scheme, the government had two strategies in mind:

(a) It increased the domestic price by 60 per cent. To put it in perspective, in 2011-12, the Indian MSP for paddy was US$0.20 per kilogram whereas the procurement price for paddy in Thailand was $0.48 per kilogram. Thai government paid more than double the price that was paid to Indian farmers.[3]

(b) The government sought to recover the money by getting higher international price. The government thought that it would be able to raise the international price by reducing Thai exports since Thailand is the second largest exporter of rice in the world. As the quantity of supply comes down, the prices would rise.[3]

Buying rice from farmers at such price proved to be ruinously expensive, costing the Thai government $12.5 billion in the first year of operation. Because of the high prices offered by the government, all farmers sold their rice to them. Criminal gangs and bent officials are said to have smuggled in thousands of tonnes of cheap grain from Cambodia and Myanmar in the hope of profiting from this policy. Storing such huge volumes also proved to be problematic. The budgetary cost of the scheme was estimated to have amounted to THB 700 billion (roughly US$22 billion), equivalent to a year of the government's total investment budget.[3, 4]

Unfortunately, after going through the trouble of procuring and storing eh food grains, the government was in for a shock. Its calculations for international market proved to be wrong. When Thailand held back its rice stocks to increase the prices, it did not count for its competitors to step in to make up for the supply shortage created by Thailand. Unfortunately for the Thais, its two closest competitors, India and Vietnam, had bumper crops that year. They took advantage of the space left empty by Thailand and exported heavily. Thus international prices did not go up for Thailand and additionally, its stocks remained unsold, instead of getting exported. Thailand is now under a mountain of rice exceeding 19 million tonnes. It is more than the government can sell domestically and export in a year.

Thai government got overthrown in 2014 and the major reason behind this was the rice pledging scheme. The government was not able to pay all the farmers for their rice at the procurement price promised. The government still owes 116 billion baht (around $3.6 billion) to about 875,900 farmers for their 6.7 million tons of pledged rice. [14, 4]

The Thailand caretaker government is struggling to raise funds to pay farmers who had pledged their rice with the government under the pledging programme. Concern is also rising over the quality of the rice piling up in the warehouses. [14, 4]

3. Universal Health Coverage Scheme

The majority of health care services in Thailand is delivered by the public sector, which includes 1,002 hospitals and 9,765 health stations. Initially, under the universal health coverage scheme, no patient paid more than Baht 30.00 (about US$1.00) per visit to accredited hospitals and health centres for inpatient or outpatient care, including drugs. In 2007, the co-payment was abolished and UCS became completely free. [25]

The Universal Coverage Scheme enrols about 74 per cent of the entire population. The scheme is financed solely from general tax revenue. Public hospitals are the main providers, covering more than 95 per cent of the insured. About 60 private hospitals joined the system and register around four per cent of the beneficiaries.

The Thai Government's financial commitment to the UCS in its first 10 years was significant. After the scheme was launched, general government expenditure on health increased from 84.5 billion baht (US$ 1.9 billion) in 2001 to 116.3 billion baht in 2002 (US$ 2.7 billion), and then it climbed steadily year on year to reach 247.7 billion baht in 2008 (US$ 7.4 billion). This was a 76 per cent real term increase from 2002, as a result of increased utilisation of health services and the rising cost of health care.

During this period, out-of-pocket payments by households decreased from 33 per cent of total health expenditure in 2001 to 15 per cent in 2008 (on a par with the average in OECD countries). Also, because there was steady growth in Thailand's GDP, total health expenditure as a percentage of GDP remained relatively stable, hovering between three and four per cent.

It has been shown that the UC Scheme has contributed significantly to reducing instances of catastrophic healthcare expenditures, especially in impoverished areas of the country. Based on the recent evaluation of the ten years of the Scheme, the number of impoverished households dropped from 3.4 per cent in 1996 to 0.8-1.3 per cent between 2006 and 2009, thus contributing to poverty reduction, building greater financial stability to vulnerable households and improved long-term livelihood security. [1]

Conclusion

Thailand never had any direct food subsidy programme. It seems to have depended mainly on income and agriculture growth for feeding its population. Other interventions also seem to have played a role - Universal health system, provision of clean water and sanitation and targeting the most vulnerable directly by old age pension scheme as well as nutrition surveillance followed by direct feeding programmes for the population at risk. As poverty levels dropped, food security was assumed to be achieved in mostly all places. When they discovered that the nutrition levels remained low, they started a detailed surveillance system to identify the malnourished and made the supplementary food reach them through school lunch programme and through village organisations.

India has school feeding programmes but it can learn from Thailand about the surveillance system that can make the food reach the needy more effectively. Mere economic growth would not lead to reduction in malnourishment in India because the process of growth has been very unequal. The fruits of the growth have not trickled down. Thus malnutrition among the poor remains high.

The Rashtriya Swasthya Bima yojana tried to do the same thing that the Thai health scheme does. However, its implementation has not received much attention of late and India still spends much lesser on the health of its poor compared to Thailand. The results are easy to see in the large difference that exists in IMR and MMR.

Since the extent of landlessness is very low among the rural population, so spread of technology was rightly thought to be adequate to achieve foods security in rural areas. Urban areas have benefited from the rapid economic growth, the problem of food security among rural households was sought to be solved by increasing the incomes of the farmers. This was attempted by increasing the procurement price by almost 50 per cent, a scheme that proved to be an economic disaster. This comes as a signal to any nation that tries to solve its poverty problems by doling out massive subsidies, without considering the economic costs involved.

References

1. Asian correspondent (2013) - "New research shows the success of Thailand's universal healthcare scheme", April 29, 2013. (http://asiancorrespondent.com/106809/new-research-continues-to-demonstrate-the-success-of-thailands-universal-coverage-healthcare-scheme/)
2. Bank of Thailand, Thailand at a glance (http://bot.or.th/English/EconomicConditions/Thai/genecon/Pages/index.aspx).
3. East Asia Forum (2014) - "Thailand's rice subsidy rotting away", 17 March, 2014 (http://eastasiaforum.org/2014/03/17/thailands-rice-subsidy-scheme-rotting-away).
4. Economist (2013) - "Thailand's Economy- the Rice Mountain" 10 August, 2013 Print Edition. (http://economist.com/news/asia/21583281-increasingly-unpopular-government-sticks-its-worst-and-most-costly-policy-rice-mountain).

5. Encyclopaedia of the nations, Asia and the pacific - Thailand poverty and wealth. (http://nationsencyclopedia.com/economies/Asia-and-the-Pacific/Thailand-POVERTY-AND-WEALTH.html).

6. International monetary fund website, data on Thailand (http://imf.org/external/pubs/ft/weo/2013/01/weodata/weorept.aspx? pr.x=84&pr.y=1&sy=2010&ey=2012&scsm=1&ssd=1& sort = country&ds=.&br=1&c=578&s=NGDP_R%2CNGDP _RPCH%2CNGDP%2CNGDPD%2 CNGDPDP C% 2CPPPPC & grp = 0&a =)

7. IRIN humanitarian news and analysis (2012) - "Thailand's rice pledging scheme", 23 August 2012. (http://irinnews.org/report/96158/thailand-rice-pledge-scheme-one-year-later)

8. Konjing K and Veerakitpanich M. (1985), "Food Consumption and Nutrition in Thailand", in: Food Policy Analysis in Thailand. Edited by Panayotou T. Agricultural Development Council, Bangkok

9. Kraisid Tontisirin, Benjawan Moaleekoonpairoj, Sakorn Dhanamitta, and Aree Valyasevi, "Formulation of supplementary infant foods at the home and village level in Thailand". http://archive.unu.edu/unupress/food/8F033e/8F033E05.htm

10. Ministry of Public Health (1984), The Royal Thai Government, "Nutrition in Primary Health Care". Bangkok.

11. National Economic and Social Development Board, Office of the Prime Minister, Bangkok, Thailand (1982), "Rural Poverty Alleviation Programme, 1982-1986".

12. NESDB (1992), 'Strengthening Food and Nutrition Surveillance for Nutrition-Oriented Development Decisions: 1991 Progress Report', UNICEF, Thailand 1992.

13. Nesdb, Economic outlook, The economic performance in 2013.

    (http://nesdb.go.th/Portals/0/eco_datas/economic/eco_state/1_56/Press%20Eng%20Q1-2013.pdf).

14. Oryza, International Journal on Rice (2012) - "Is rice pledging scheme throwing thailand into a maze of rules?", 27 February, 2012. (http://oryza.com/news/rice-news/rice-pledging-scheme-throwing-thailand-government-maze-rules).

15. Rappler social news network (2014) - "What Philippines can learn

from Thai school nutrition programs", 2 December, 2014. (http://www.rappler.com/move-ph/issues/hunger/50365-thailand-school-nutrition-programmes)

16. Royal Thai Embassy website article on Millennium development Goals. (http://thaiembdc.org/dcdp/node/588)

17. Suwanrada, W. and Kamwachirapithak, A. (2007). "Current Situation Analysis of the Old-Age Allowances System in Thailand". Research Report submitted to Thailand Research Fund (Project No. PDG5030208), August 2007.

18. Teller C. H. (1987), "Community Nutrition Assessment and Evaluation: Towards Model Building", Thailand Trip Report Aug 15-29, 1987; International Nutrition unit Technical Report Series, U.S. Department of Health and Human Services, 1987.

19. Thainews website (2012) - "Thailand's position in ASEAN", 11 December, 2012. (http://thainews.prd.go.th/centerweb/newsen/News Detail? NT01_News ID = WNSOC 5703270010003)

20. Thomas white International- "Thailand: Asian Tiger leads regional Growth". (http://thomaswhite.com/world-markets/thailand-asian-tiger-leads-regional-growth/)

21. Unakul S (1985), "An Economic Planner's View of the Nutrition Problem in Thailand", in: Food Policy Analysis in Thailand. Edited by Panayotou T. Agricultural Development Council, Bangkok

22. United Nations Development Program in Thailand- (http://th.undp.org/content/dam/thailand/docs/TICAUNDPbpVol1.pdf)

23. Wikipedia article on History of Thailand since 1973. (http://en.wikipedia.org/wiki/History_of_Thailand_since_1973)

24. Wikipedia article on Thailand. http://en.wikipedia.org/wiki/Thailand

25. Wikipedia article on health in Thailand. (http://wikipedia.org/wiki/Health_in_Thailand)

26. Winichagoon P and Schuftan C (1991), "Thailand: Food and Nutrition Surveillance System: An Inquiry into Output Data Utilisation for Decision-Making", 9-15 December 1991.

27. World Bank website, overview of Thailand including a plethora of data. (http://worldbank.org/en/country/thailand/overview)

PART III
NATIONS WITH COMPARABLE FOOD SECURITY

7

Nepal

Compilation: SaharshYashlaha

Introduction

Nepal, officially the Federal Democratic Republic of Nepal, is a landlocked sovereign state located in South Asia. With an area of 147,181 square kilometers (56,827 sq mi) and a population of approximately 27 million (and nearly two million absentee workers living abroad), Nepal is the world's 93rd largest country by land mass and the 41st most populous country. [8]

Nepal is an agrarian-based economy. Agriculture provides livelihood for three-fourth of the population. It accounts for 38 per cent of GDP but low agricultural productivity and traditional methods of farming have left more than 33 districts food deficient in Nepal. It is estimated that more than 3.7 million people in Nepal are at the risk of chronic hunger. Nepal's dependency on imported food grains has significantly grown over the past few years. Food production has gone down and made Nepal dependent on imports. As per the Global Hunger Index-2012, Nepal has slid to 60th position from 54th in 2011. Seventeen per cent of the country's population is malnourished. Similarly, more than 38 per cent of the children below five years of age are stunted, and five per cent of these stunted children die before they reach five years.

Additional challenges to Nepal's growth include its landlocked geographic location, persistent power shortages, underdeveloped transportation infrastructure, civil strife and labour unrest, and its susceptibility to natural disaster. The lack of political consensus in the past several years has delayed national budgets and prevented much-needed economic reform. [2]

2) Indicators for food Security

2.1) Economic Indicators

2.1.1) Real Gross Domestic Product Growth Rate:

Definition of GDP-real growth rate: This entry gives GDP growth on an annual basis adjusted for inflation and expressed as a per cent.

Table 1 shows that the growth rates of GDP for all the years mentioned are consistently higher than those of Nepal. India is developing at much faster rates than Nepal.

Table 1: Real Growth Rate of Nepal and India

Year	1999	2000	2002	2003	2004	2005	2006	2007	2008	2009	2010	2011
Nepal	3.4	3.7	-0.6	3	3	2.7	1.9	3.2	4.7	4.7	4.6	3.5
India	5.5	6	4.3	8.3	6.2	8.4	9.2	9	7.4	7.4	10.4	7.2

*[**Source:** CIA World Factbook, 2011(http://www.indexmundi.com/g/g.aspx?c = np &v = 66)]*

Nepal Inflation Rate Mean: 7.4727
India Inflation Rate Mean: 6.6766
Nepal Standard Deviation Inflation Rate: 3.4378
India Standard Deviation Inflation Rate: 2.6341

Paired T-test

t-Test: Paired Two Sample for Means

	Variable 1	*Variable 2*
Mean	3.15	7.441667
Variance	2.115455	3.022652
Observati	12	12
Pearson C	0.434468	
Hypothesi	0	
df	11	
t Stat	-8.66925	
P(T<=t) or	1.51E-06	
t Critical o	1.795885	
P(T<=t) tw	3.02E-06	
t Critical t	2.200985	

(2.1.2) Inflation Rate

Both the nations India and Nepal have been facing the situation of stagflation since 2008. The onset of stagflation seems to be a delayed by one year for Nepal but both countries saw near double digit inflation and slow growth in this period. The inflation rates are slightly higher and seem to be fluctuating a little more widely in Nepal (the average inflation rate is 7.47 per cent and the standard deviation is 3.44) compared to India (the mean inflation rate is 6.68 and the standard deviation is 2.63). But the paired t test shows that the two are not statistically significantly different (t- statistic = 0.525 and p-value = 0.305).

Both nations thus suffer from high inflation rates but India's growth compensates for its inflation better than Nepal's.

Table 2: Inflation Rate of Nepal

Country	1998	1999	2001	2002	2005	2006	2007	2008	2009	2010	2011
Nepal	11.8	3.3	2.8	2.9	7.8	8.6	6.4	7.7	13.2	8.6	9.1

*[**Source:** CIA World Factbook, 2011(http://www.indexmundi.com/g/g.aspx?v =71&c =n p&l = en)]*

Table 3: Inflation Rate of India

Country	1999	2000	2002	2003	2004	2005	2006	2007	2008	2009	2010	2011
India	6.7	5.4	5.4	3.8	4.2	4.2	5.3	6.4	8.3	10.9	11.7	8.9

*[**Source:** CIA World Factbook, 2011(http://www.indexmundi.com/g/g.aspx?c = np&v =71)]*

(2.2) Demographics and Social Indicators

Health Indicators

Infant Mortality Rate in Nepal has decreased from 74.14 in 2001 to 43.13 in 2012 while it decreased from 63.19 to 46.07 in India for the same period as shown by Tables 4 and 5 respectively for Nepal and India. We used the semi log regression to estimate the rate of decline.

Log (IMR) = a + b (Year) where b directly gives the rate of growth.

The regression shows that the IMR decreased at the rate of 5.4 per cent in Nepal (R square = 0.885 and t statistic = 8.8) while it declined at much slower rate, i.e., at 2.85 per cent (R square = 0.99 and t statistic = 97.34) in India for the same period.

Similar is the case for MMR, except that here the decline in MMR was much faster than IMR for Nepal (Table 6). Compared to India, MMR was 1.5 times more for Nepal in 2000; it decreased considerably and stood at exactly the same level as India in 2012.

The literacy rates have also increased very fast over the last decade for Nepal: it started from mere 20 per cent literacy but reached 60 per cent by 2012 much closer to India's 72 per cent. Thus for the health indicators, Nepal used to be below India but with its fast progress, it has caught up with India in MMR and is slightly ahead in IMR.

Table 4: Infant Mortality Rate of Nepal

Country	2001	2002	2003	2004	2005	2006	2007	2008	2009	2010	2011	2012
Nepal	74.14	72.36	70.57	68.7	66.9	65.3	63.7	62	47.5	46	44.5	43.1

*[**Source:** CIA World Factbook, 2011(http://www.indexmundi.com/g/g. aspx ?c = np & v = 29)]*

Table 5: Infant Mortality Rate of India

Country	2001	2002	2003	2004	2005	2006	2007	2008	2009	2010	2011	2012
India	63.19	61.47	59.59	57.92	56.29	54.63	53	52	50	49.13	47.57	46.07

*[**Source:** CIA World Factbook, 2011(http://www.indexmundi.com/g/g.aspx ? v = 29 & c = in & l = en)]*

Table 6: Maternal Mortality Rate of Nepal and India

Year	1990	1995	2000	2005	2013
Nepal	790	580	430	310	190
India	560	460	370	280	190

*[**Source:** CIA World Factbook, 2014(http://www.who.int/gho/maternal_health/ countries/ npl. pdf? ua =1)]*

Table 7: Literacy Rate of Nepal

Year	1981	1991	2011	2010
Literacy Rate	20.57	32.98	48.61	60.31

*[**Source:** CIA World Factbook(http://www.indexmundi. com/facts/nepal/liter acy-rate)]*

India

Table 8: Literacy Rate of India.

Year	1995	2001	2003	2006	2011
Literacy Rate	52	61	59.5	62.75	74.04

*[**Source:** CIA World Factbook (http://www.indexmundi.com/g/g.aspx ?v = 39 & c =ce& l=en)]*

Malnutrition Indicators

The proportion of undernourished population is almost the same for both the nations (Table 11). However the percentage of underweight children in Nepal (Table 12) is lesser than India. The Global hunger index (Table 13) score has been 4 points lower for Nepal. Thus Nepal has little lesser incidence of hunger. It is interesting to observe that a nation lagging behind in all growth indicators should be doing better, even if slightly, than India in dealing with the malnutrition problem.

In overall performance of HDI also (Tables 9 and 10), Nepal is catching up with India. However, at present India is statistically significantly higher than Nepal in HDI mainly because of its emerging economy. India's performance is much less impressive in the social sectors.

Table 9: HDI Value of Nepal

	Life Expectancy at Birth	Expected Years of Schooling	Mean Years of Schooling	GNI per capita (2005 PPP$)	HDI value
1980	48.2	4.5	.6	0,566	0.234
1985	51.1	5.5	1.2	0,633	0.285
1990	54	7.4	2	0,706	0.341
1995	57.5	8	2.2	0,811	0.370

2000	61.6	8.8	2.4	0,902	0.401
2005	65.6	8.9	2.7	0,960	0.429
2010	68.5	8.9	3.2	1,090	0.458
2011	68.8	8.9	3.2	1,105	0.460
2012	69.1	8.9	3.2	1,137	0.463

*[**Source:** Human Development Report, 2013 (http://hdr.undp.org/sites/default/files/Country-Profiles/NPL.pdf)]*

Table 10: HDI Value of India

	Life Expectancy at Birth	Expected Years of Schooling	Mean Years of Schooling	GNI per capita (2005 PPP$)	HDI value
1980	55.3	6.3	1.9	0,880	0.345
1985	57	7.1	2.4	1,007	0.379
1990	58.3	7.4	3	1,191	0.410
1995	59.8	8.2	3.3	1,389	0.438
2000	61.6	8.3	3.6	1,702	0.463
2005	63.3	9.9	4	2,190	0.507
2010	65.1	10.7	4.4	3,009	0.547
2011	65.4	10.7	4.4	3,175	0.551
2012	65.8	10.7	4.4	3,285	0.554

*[**Source:** Human Development Report, 2013 (http://hdr.undp.org/sites/default/files/Country-Profiles/IND.pdf)]*

Table 11: Undernourishment in Nepal and India

Year	1990-92	1994-96	1999-01	2004-06	2010-12
Nepal	25.9	27.1	24.5	21.7	18.0
India	26.9	25.2	21.3	20.9	17.5

*(**Source:** Global Hunger Index Data from IFPRI)*

Table 12: Underweight Children in Nepal and India

Year	1988-92	1993-97	1998-02	2003-07	2008-12
Nepal	44.6	44.1	43	38.8	29.1
India	59.5	45.9	44.4	43.5	40.2

*(**Source:** Global Hunger Index Data from IFPRI)*

Table 13: GHI of Nepal and India.

Year	1990	1995	2000	2005	2013
Nepal	28.0	27.3	25.3	22.3	17.3
India	32.6	27.1	24.8	24.0	21.3

*(**Source:** Global Hunger Index Data from IFPRI)*

(3) Economic and Political History

An isolated, agrarian society until the mid-20th century, Nepal entered the modern era in 1951 without schools, hospitals, roads, telecommunications, electric power, industry, or a civil service. The country has made progress towards economic and social growth since then. But Nepal even today ranks among the world's poorest countries, with a per capita income of around $490 in 2009. In 2008, it was estimated that 25 per cent of the population was below the poverty line.

The country receives substantial amounts of external assistance from India, the United Kingdom, the United States, Japan, and the European Union (EU). Several multilateral organisations–including the World Bank, the Asian Development Bank, and the UN Development Programme–also provide significant assistance. [3]

Agriculture remains Nepal's principal economic activity; employing around 70 per cent of the population and providing about one-third of GDP. Only about 25 per cent of the total area is cultivable; another 33 per cent is forested; most of the rest is mountainous. Rice and wheat are the main food crops. The lowland Tarai region produces an agricultural surplus that it makes available to the food-deficient hill areas. [7]

(3.1) 1768-1950s: [9]

For more than two centuries since the creation of modern Nepal in 1768, the economic system was feudal. Preoccupation with territorial expansion through military conquests defined the present boundaries of the country in the early 19th century. But such conquests were without much thought for the country's economic well-being. The ultimate economic burden of war and such conquests fell on the poor peasantry who not only supplied manpower to the war but also fed and paid for war machine. This Rana period (Nepal was at that time ruled by Jang Bahadur Rana) witnessed family enrichment to unprecedented level. It was made possible due to:

(a) Ruthless exploitation of land grant system to fill personal coffers; and

(b) Use of compulsory labour in the form of Jhara and Rakam to build lavish palaces.

The post Rana decade of the 1950s saw unprecedented popular awareness. The decolonisation movement world-wide and successful anti-Rana struggle brought unprecedented awareness of liberal and modern values among the people.

The general election in 1959 brought to power a socialist government under a charismatic leadership of B. P. Koirala. But political instability, shrewd manoeuvrings by the monarch to expand his space and eventual royal takeover in December 1960 from a popularly elected brought the democracy to an abrupt end.

(3.2) Panchayat Raj (1960-1990): [9]

Interestingly, it is believed that the absence of political dissention with unquestioned loyalty to the monarch ensured political stability. This stability enabled the government to concentrate on developmental issues. A review of the principal economic achievements during this period shows an annual average GDP growth of 3.85 per cent and per capita income growth of 1.5 per cent (this is similar to India's growth rates of the same period). The dependence on agriculture came down: Share of agricultural GDP declined from 65 per cent to 51 per cent. Improvements were made during 1965-89 period in socio economic indicators: life expectancy at birth increased from 41 to 59, infant mortality fell 171 to 124, daily calorie intake increased marginally from 1887 to 2078. The most dramatic change was experienced in the education sector: female enrolment jumped from 4 to 57 per cent and primary school enrolment went up from 20 to 86 per cent. However, political opposition to the royal regime reached its crescendo in 1990 which brought an end to the Panchayat regime.

(3.3) Democratic Period: [9]

The Nepali Congress was voted to power for most of the post Panchayat period of the 1990s and early part of the new millennium. Although the period was not free from political instability caused by frequent changes in government and eventually the royal takeover, the period witnessed significant accomplishments. Some of them included:

- GDP growth of 4.95 until 2001. Per capita income growth of two per cent.
- Share of agriculture to GDP went down from around 50 per cent in early 1990s to 39 per cent in 2002-03. Population depending on agriculture also went down from 81 per cent to 66 per cent.
- Poverty declined from 42 per cent to 31 per cent during 1995/6 to 2003/4.
- Literacy improved from 40 to 54 per cent, and life expectancy increased to 61 years.

(3.4) Why Malnutrition Rates are High [4]

(3.4.1) Population Growth

In Nepal population growth has exceeded the long-term growth in income and other factors that contribute to poverty reduction, thereby contributing to the vicious circle of poverty. Although real growth in the economy exceeded population growth in the 1970s, 1980s and the 1990s, growth in the agriculture sector on which the livelihoods of the majority of rural people are dependent, has lagged behind population growth in both the 1970s and the 1990s.

Table 14: Comparative Growth in Population and Economy

	1971-81	1981-91	1991-2001
Population Growth	2.66	2.10	2.28
Real GDP Growth	3.11	5.06	4.64

(3.4.2) Limited Scope for Expansion of Cultivated Areas

The interacting problems of people's heavy dependence on agriculture and higher growth of farm population than total population are compounded by extremely limited scope for expansion of cultivated area without heavy environmental cost. This is the second factor contributing to accentuation of poverty and the possibility for further deterioration of poverty and food insecurity in the future. It is significant that agricultural land, which increased by about 767 000 ha during the 1970s, increased only by 34 000 ha in the 1980s.

(3.4.3) Slow and Variable Growth in Agriculture and Low Productivity

Another phenomenon that helps account for the growing problems of poverty and food insecurity is the slow and variable growth of the dominant agriculture sector. The study by FAO presented in a report Nepal Agricultural Policy and Strategies for Poverty Alleviation and Food Security shows that although the cereal production grew by 2.11 per cent during the 1990s, the growth in cereal yields was only1.20 per cent. The production fluctuates widely from year to year; thus there is a lot of risk involved for farmers.

(3.4.4) Unequal Land Distribution

The land distribution has been highly skewed in favour of big landholders in Nepal. Same study by FAO presents the Gini Coefficients for four consecutive agricultural census periods that are reproduced here. The GC is a measure of inequality: the higher the coefficient (maximum 1), the greater the degree of inequality. Probably due to the cumulative effect of land reform measures adopted in the 1960s, by 1980s, the situation started improving; the Gini declined notably in 1991-92.

Table 15: Gini Coefficient.

Census Year	1961-62	1971-72	1981-82	1991-92
Real GDP Growth	0.82241	0.83614	0.80257	0.68998

Above calculations in table are done by FAO in a Report Nepalagricultural Policy and Strategiesfor Poverty Alleviation and Food Security. Source: ftp://ftp.fao.org/docrep/fao/008/ae898e/ae898e06.pdf

(3.4.5) Nepal's Efforts Against Malnutrition [6]

Nepal has achieved spectacular success in reducing child mortality rate over last decade as shown earlier in Table-4. But Nepal still has a long way to go: The Nepal Demographic and Health Survey (NDHS) PEM (Protein Energy Malnutrition) 2006 showed that 49 per cent of children aged less than five years were affected by stunting, 39 per cent were underweight, and 13 per cent were wasted.When compared with the NDHS 2001, some improvements have been noted: The percentage of children stunted fell by 14 percentage points from 57 per cent in 2001 to 49 per cent in 2006. Underweight has declined only slightly (from 43 per cent to 39 per cent), and wasting has increased from 11 per cent in 2001 to 13 per cent in 2006. One of the aims of MDG (Millennium Development Goals) in Nepal is to reduce the prevalence of malnourishment in children of age less than five year from 57 per cent in 1990 to 29 per cent in 2015. However, the above results indicate that, achieving this MDG of a 50 per cent reduction would be a challenge.

(3.4.6) Government's Five Year Plans [5]

Until the Fifth Plan (1975-80) accelerating economic growth remained the primary objective of development planning in Nepal. Poverty reduction as a goal was implied because of the expected trickle- down effect of rapid economic growth. Experiences over the period, however, showed that the trickle-down effect did not work beyond the urban areas. Even after 25 years of planned efforts, 36 per cent of population was living in poverty. Recognising this, the Sixth Plan in 1980 explicitly included poverty reduction as one of its goals, yet poverty kept on increasing, growing to 42.5 per cent in 1985. The Seventh Plan then formulated a separate programme termed the 'Basic Needs Fulfilment Program' for poverty alleviation. The objective was to eliminate poverty within 15 years, but the Program did not succeed and was abandoned following the political changes of 1990. Against the background of the restoration of multiparty democracy, the Eighth Plan (1992-97) was formulated with poverty reduction as one of the main objectives. This Plan also failed to make any notable progress in reducing poverty.

(4) Current Policies

(4.1) Public Distribution System [1]

There is only limited public food distribution system in Nepal. At present mid-day meals are being provided to primary school age children under the WFP funding programme (discussed in more details under the second scheme). This is a much appreciated programme but its coverage is very limited. The programme is also unsustainable as government's own resources are not enough to continue this programme. Once the World Food Program withdraws food support to this programme, it would be most probably discontinued.

At present there is no food distribution programme for pregnant and lactating mother. The coverage of public distribution system is also narrowing because most of the food depots in remote areas had to be closed due to insecurity. There were instances of insurgents looting the food stock available at some food depots. The coverage of food distribution system is also shrinking because of lack of funds.

Nepal's public food distribution system was never a significant operation and hardly had any impact on overall food security situation of the country. It, however, did contribute to some extent to food security of the people living in the remote areas of Nepal. Public distribution system of food in Nepal has also been criticised as a programme that ensures food security of government officials posted in remoteregions rather than the needy local populace.

(4.2) Country Programme (2013-2017)[10]

WFP food assistance provided under the Country Programme aims to improve the nutritional status of women and children and to encourage children to go to school. The programme is operative in 14 districts of mid and far western regions of Nepal. The programme consists of two main activities:

Activity 1: Food for Education

By providing a midday meal and deworming medication, the project improves attendance rates and the nutritional health of school-going children in food insecure areas. Where large disparities exist in attendance rates between girls and boys, a monthly ration of cooking oil is given to the mothers to encourage families to send their daughters to school.

Activity 2: Mother and Child Health Care

With nearly every other pregnant woman and child under five anemic in Nepal, this programme addresses micro-nutrient deficiencies and malnutrition through monthly distributions of take home rations of fortified food. As a part of the programme, women receive pre- and postnatal health care and the children's growth is monitored. Information on nutrition and good healthcare practices are also shared with the mothers.

(5) Conclusion

Nepal has been experiencing a lot of political instability and hence it is not surprising that it does not have any substantial programme for ensuring food security. Yet, the nation's health indicators as well as levels of malnutrition are not significantly higher than India's. With much more resources put in by India, the achievement in food security levels has not been any better than the nation which has put hardly any efforts. Nepal's indicators seem to have grown totally based on its economic prosperity; political activities have hindered the progress of social sector indicators but have not been able to stop the betterment process of the population. It raises an interesting question for India: if we stop all our welfare expenditures and concentrate totally on growth, would we still be able to reduce our malnutrition rates? However, knowing that a large proportion of population still derives some relief from the PDS programme, we should look at making our schemes more effective rather than abandoning them. If we invest our resources more effectively, we would surely be able to see more progress than nations like Nepal.

References

1. BalGopalBaidya, "Food Security Situation in Nepal" http://adrf.trf.or.th/ADRF 6update/Full_ Papers/Food_Security Bal_ Baidya/ Full Paper_Bal_Baidya. pdf
2. Central Intelligence Agency (CIA)The World Factbook, 2014 https://www.cia.gov/library/publications/the-world-factbook/geos/np.html
3. Central Intelligence Agency (CIA) The World Factbook, "Nepal Economy", 2012
http://globaledge.msu.edu/countries/nepal/economy
4. FAO (Food and Agriculture Organisation), "Economy, Poverty and Food Security ", 2003
ftp://ftp.fao.org/docrep/fao/008/ae898e/ae898e06.pdf
5. FAO(Food and Agriculture Organisation), "The Macro Policy and Institutional Environment", 2003
ftp://ftp.fao.org/docrep/fao/008/ae898e/ae898e07.pdf
6. GaganGurung, "Social Determinants of Protein-Energy Malnutrition: Need to Attack the Causes of the Causes", 2010
http://www.jstor.org/stable/pdfplus/10.2307 23499827.pdf? acceptTC = true&__redirected
7. Invest Nepal, "About Nepal", 2014
http://www.investnepal.gov.np/portalindex. php? p1 = content & p2 = 5#. U8YjQPmSyPU
8. PramitaAryal, "Right To food", 2014
http://trn.gorkhapatraonline.com/index.php/op-ed/4089-right-to food-pramita-aryal.html
9. Ram S Mahat, "Economic side of Nepal's political economy"
http://www.ramsmahat.com/articles Political % 20 Economy % 20 of%20Nepal%27s%20Development%28Revised%29.doc
10. WFP (World Food Programme),"Food Security Atlas for Nepal"
http://www.foodsecurityatlas.org/npl/country

8

Bangladesh

Compilation: Manas Roy

Introduction

Bangladesh is located in South Asia bounded by India in the West, North, and East, the Bay of Bengal in the South and Myanmar in the Southeast. It has a 4,685 km long boundary and has a unique geographical location in South Asia. It is situated at the lower part of the Ganges-Brahmaputra- Meghna (GBM) basin that emanates mostly from the Himalayas and is home to these three mighty rivers. Over 200 rivers and tributaries have made this country a land of rivers. These rivers originate in the great Himalayas, flow through Bangladesh and flow into the Bay of Bengal from there.

Agricultural productivity has increased in Bangladesh in the last two decades and it is close to achieving self-sufficiency in food grain (rice and wheat) production. Since independence in 1971, Bangladesh has also achieved substantial improvements in some social indicators like decreasing infant and maternal mortality as well as illiteracy, increasing life expectancy, access to safe water and sanitation. The country's economy has grown Five-Six per cent per year since 1996 [3].

In spite of this, Bangladesh faces high poverty and child under-nutrition rates, which are exacerbated by frequent natural disasters and high population density. Despite important economic progress, the country remains highly food-insecure. Other development indicators are also low: Bangladesh is ranked 146th out of 186 countries in the 2013 Human Development Index (HDI) and 68th out of 79 countries in the 2012 Global Hunger Index (GHI). According to the WFP Household Food Security and Nutrition Assessment (2008-2009), 37 million people – a quarter of the population are food insecure [17].

Bangladesh has made progress in terms of net enrolment rates and has achieved gender parity in primary and lower secondary education. Major problems remain in terms of school completion, particularly for girls from extremely poor families. An estimated 3.3 million children remain out of school, and only 51 per cent of those in school complete the full five-year cycle of primary education (UNICEF 2009).

The Government's Poverty Reduction Strategy continues to address these challenges through the National Food Policy (2006) and Plan of Action (2008-2015). These programmes focus on the following goals: Stable supply of nutritious food and increased purchasing power for the people; and adequate nutrition for all individuals especially women and children. These goals will be achieved through intensifying investment in agricultural research and livestock diversification, strengthening rural population's access to markets and women's access to productive assets and expanding social safety net programmes [26].

Economic and Demographic Indicators

Literacy rates have doubled in Bangladesh between 1981 and 2011 from 29 per cent to 58 per cent (Table 1) while it has increased by 80 per cent in India from 41 per cent to 78 per cent. It is to be noted that education expenditure of India was 3.3 per cent of GDP in 2010 while of Bangladesh was 2.2 per cent of GDP in 2009. Thus the policies may seem to be more effective in Bangladesh. However, the literacy rate of India is 20 per cent points higher than that of Bangladesh this in spite of the fact that India has a much larger population to cater to compared to Bangladesh. In other words, India has been more effective than Bangladesh in educating its population.

The growth rate of Bangladesh (average for 2001 to 2011 is 5,89) is lower than that of India (average growth for the last decade is 7.18) but the growth has been more steady compared to India – the standard deviation for Bangladesh is 0.72 whereas it is 2.28 for India (Table 2.1). India has been growing faster than any of its counterparts in South Asia. But Bangladesh is not too far behind in growth rates.

Average inflation rates in both the nations are comparable 6.75 for Bangladesh and 6.06 for India. (Table 3.1); the rates of fluctuations also are similar (2.69 for Bangladesh and 2.56 for India).

The percentage of population living below the poverty line is also slightly lower for India 32 per cent of the population in 2005 was living below the poverty line whereas 40.6 per cent were poor in Bangladesh (Table 5). However, while poverty declined by about 41 per cent in Bangladesh between 1992 and 2010, it declined in India by about 30 per cent. Although the proportion of poor population remains higher in Bangladesh and the rate of economic growth was lower than India, the rate of fall in population living below the poverty line was faster in Bangladesh. The explanation can be found in the reason given below by the Bangladesh Poverty Assessment the growth in incomes was for the labour in Bangladesh while it went more to the rich in India.

The Bangladesh Poverty Assessment shows that during the period 2000-2010 poverty reduction was closely linked to the growth in labour income and changes in demographics. Economic growth led to rise in labour income, in both formal and informal sector. That led to lower poverty rates. Parallel to this, fertility rates (including those of the poor) have been steadily dropping over the last several decades which have resulted in lower dependency ratios thereby increasing income per capita and reducing poverty. The proportion of the population living under the poverty line came down from 49 per cent in 2000 to 34.5 per cent in 2010 as a result of consistent economic and remittance growth (Household Income and Expenditure Survey HIES 2010). Poverty declined in Bangladesh 1.8 per cent annually between 2000 and 2005, and 1.7 per cent annually (est.)

between 2005-2010. This rate of decline is a positive indication towards achieving the MDG target of poverty reduction [17].

Compared to Bangladesh, India spends more on its population's welfare 62.7 per cent of the public expenditure was spent on welfare schemes in India. Compared to that, Bangladesh spends 38.8 per cent (Table 4). The cost effectiveness of the expenditure thus might be more for Bangladesh.

Table 1: Literacy Rate

Country			Years		
	1971	1981	1991	2001	2011
Bangladesh	16.8	29	35.3	47.5	58
India	34.45	41	48	61	78

***(Source:** World bank social indicators) http://data.worldbank.org/indicator/SE.ADT.LITR.ZS/countries/BD?display=default)*

Table 2: GDP Growth Rate (Annual Per cent)

Country	Year(s)										
	1987	1995	2001	2002	2003	2004	2005	2006	2007	2008	2009	2011
Bangladesh	3.7	4.9	5.3	4.4	5.3	6.3	6	6.6	6.4	6.2	5.7	6.7
India	4	7.6	4.8	3.8	7.9	7.9	9.3	9.3	9.8	3.9	8.5	6.6

***(Source:** World Bank Social Indicators) http://data.worldbank.org/indicator/NY.GDP.MKTP.KD.ZG/countries/BD?display=default)*

Table 2.1: Mean and Standard Deviation of Growth Rates for Bangladesh and India.

Bangladesh: Average rate of the growth (2001 to 2011): 5.89. Standard Deviation: 0.72

India: Average rate of the growth (2001 to 2011): 7.18. Standard Deviation: 2.28

Table 3: Inflation Consumer Prices

Country	Year(s)											
	1987	1995	2001	2002	2003	2004	2005	2006	2007	2008	2009	2011
Bangladesh	9.9	10.3	2	3.3	5.7	7.6	7	7.8	9.1	8.9	5.4	10.7
India	8.8	9.4	3.7	4.4	3.8	3.8	4.2	6.1	6.4	8.4	10.9	8.9

***(Source:** World Bank Social Indicators)* http://data.worldbank.org/indicator/FP.CPI.TOTL.ZG/countries/BD?display=default

Table 3.1: Mean and Standard Deviation of Inflation for Bangladesh and India

Bangladesh: Average rate of the inflation (2001 to 2011): 6.75. Standard Deviation: 2.69

India: Average rate of the inflation (2001 to 2011): 6.06. Standard Deviation: 2.56

Table 4 : Subsidies and Other Transfers (as Per cent of Expense)

Country	Year(s)								
	2002	2003	2004	2005	2006	2007	2008	2009	2011
Bangladesh	25.3	28.5	29.5	30.6	29	29.3	31.5	29.9	38.8
India	42.6	49.3	46.8	49.2	53.9	59.7	61.4	60.1	62.7

Source: *World Bank Social Indicators* http://data.worldbank.org/indicator/GC.XPN.TRFT.ZS

Table 5: Population Living Below Poverty Line

Country	Year(s)				
	1992	1995	2000	2005	2010 est.
Bangladesh	58.8	51	49.8	40.6	34.5
India	42	41	36	32	29.8

*(**Source:** World Bank Social Indicators http://data.worldbank.org/indicator/SI.POV.NAHC/countries)*

Table 6 : Per Capita Availability of Food Grains (in kgs Per annum)

Country	Years (s)										
	1991	1995	2001	2002	2003	2004	2005	2006	2007	2008	2009
Bangladesh	165	165.5	188.30				192.23				
India	186.2	180.8	151.9	180.4	159.7	168.9	154.2	162.5	161.6	159.1	162

*(**Source:** planning commission of India planningcommission.gov.in/sectors/agri_html/selagri/T1.25.xls)*

Malnutrition Indicators in Bangladesh

Per capita GDP doubled between 1990 and 2010, and agricultural growth averaged 3.3 per cent per year due to impressive gains in rice yields [2].

As the figures show, both India and Bangladesh have alarmingly high rates of malnourishment. 42 per cent of the children are malnourished in India; the figure is just a little lower at 36 per cent for Bangladesh (Table 7). The percentage of stunted children is similar at 43 per cent and 41 per cent respectively for India and Bangladesh.

There is not much difference in the mortality rates either: IMR is 5 points higher in India, under-5 mortality rate is 14 points higher and MMR shows 20 more deaths per 100,000 live births for India (tables 10, 11 and 12). However, data from different sources show different figures. Therefore we do not know the accuracy of the data. The only conclusion that we can draw is that the difference between the indicators seems to be only marginal. It looks like both the nations are at equally low levels. Even the percentage of GDP spent on health expenditures is almost similar - 3.6 per cent in Bangladesh and 4 per cent in India (Table 9), thus the experiences of the nations are similar. Both have a long way to go in implementing their policies more effectively.

Table 7: A Comparison of Under 5 Malnourished Children (Per cent) in India and Bangladesh

Country	Year(s)			
	1990	1997	2006	2011
Bangladesh	61.5	52.5	39.8	36
India	59.5	41.1	43.5	42

Source- *World Bank social Indicators, The Hindu, Sources- UNESCO report on Bangladesh, Reuters article*

http://data.worldbank.org/indicator/SH.STA.STNT.ZS/countries

Table 8: A Comparison of Under 5 Stunted Children (Per cent) in India and Bangladesh

Country	Year(s)		
	1996-97	2007	2011
Bangladesh	60	43	41
India	52 (1992-93)	48 (2004)	43

*(**Source**- World Bank social Indicators, The Hindu, Sources- UNESCO report on Bangladesh, Reuters article*

http://data.worldbank.org/indicator/SH.STA.STNT.ZS/countries)

Table 9 : A Comparison of Expenditure of Government on Health (as Per cent of GDP)

Country	Year (s)			
	1995	2000	2005	2012
Bangladesh	3.5	2.8	3.1	3.6
India	4	4.3	4.5	4

*(**Source-** World Bank social Indicators, The Hindu, Sources- UNESCO report on Bangladesh, Reuters article*

http://data.worldbank.org/indicator/SH.STA.STNT.ZS/countries)

Table 10: A Comparison of IMR Between India and Bangladesh

Country	Year(s)					
	1980-1985	1985-1990	1990-1995	1995-2000	2000-2005	2005-2010
Bangladesh	122.5	104.4	90.6	79.8	59.3	49.0
India	105.5	88.5	83	77	63	54

*(**Sources** World Bank social Indicatorshttp://data.worldbank.org/indicator/SP.DYN.IMRT.IN/countries)*

Table 11: A Comparison of Under 5 Mortality Rate in India and Bangladesh

Country	Year(s)			
	1980	1990	2000	2010
Bangladesh	198	144	88	47
India	166	126	92	61

*(**Source**: World Bank social Indicatorshttp://data.worldbank.org/indicator/SH.DYN.MORT/countries)*

Table 12: Maternal Mortality Rate Modelled Estimate Per 100,000 live Birth

Country		Year(s)		
	1990	1995	2000	2010
Bangladesh	550	440	340	200
India	560	460	370	220

*(**Source**: World Bank social Indicators http://data.worldbank.org/indicator/SH.STA.MMRT/countries)*

Economic History

The partition of India and the emergence of Pakistan in 1947 severely disrupted the economic system in East Pakistan. The united government of Pakistan expanded the cultivated area and some irrigation facilities, but the rural population generally became poorer between 1947 and 1971 because improvements did not keep pace with rural population increase. Pakistan's five-year plans opted for a development strategy based on industrialisation, but the major share of the development budget went to West Pakistan, that is, contemporary Pakistan. The lack of natural resources meant that East Pakistan was heavily dependent on imports, creating a balance of payments problem. Without a substantial industrialisation programme or adequate agrarian expansion, the economy of East Pakistan steadily declined

After 1971, when Bangladesh was formed, there were critical shortages of essential food grains and other staples because of wartime disruptions. External markets for jute had been lost because of the instability of supply and the increasing popularity of synthetic substitutes. Foreign exchange resources were minuscule, and the banking and monetary systems were unreliable. Although Bangladesh had a large work force, the vast reserves of under trained and underpaid workers were largely illiterate, unskilled, and underemployed. Commercially exploitable industrial resources, except for natural gas, were lacking. Inflation, especially for essential consumer goods, ran between 300 and 400 per cent. The war of independence had crippled the transportation system. Hundreds of road and railroad bridges had been destroyed or damaged, and rolling stock was inadequate and in poor repair. India helped with critically measured economic assistance in the first months after Bangladesh achieved independence from Pakistan.

After 1975, Bangladeshi leaders began to turn their attention to developing new industrial capacity and rehabilitating its economy. The static economic model adopted by these early leaders, however-including the nationalisation of much of the industrial sector-resulted in inefficiency and economic stagnation. Beginning in late 1975, the government gradually gave greater scope to private sector participation in the economy, a pattern that has continued. Many state-owned enterprises have been privatised, like banking, communication, jute, and media. Such liberalisation led to higher growth in the economy [14].

Bangladesh government began to liberalise agriculture extensively since the early 1990s. Prior to that, plant-breeding research and infrastructure development were accelerated throughout the 1970s and the 1980s. Both agricultural inputs and output markets were liberalised between 1988 and 1993; private imports of fertiliser and both grains (rice and wheat) were legalised in 1993. The liberalisation of the fertiliser and minor irrigation equipment occurred during 1988-1992.Treating 1992–93 as the start of liberalisation, the post-liberalisation period has witnessed acceleration in the growth of supply as well as production.

The Bangladesh economy has made respectable progress in rice, tripping production from 11 million tonnes in 1971 to 33 million in 2012. The growth of production was achieved by fast adoption by farmers of higher yielding crop which was supported by rapid expansion of irrigation infrastructure through private investment in tube wells. Increased domestic food grain production, lower real prices of rice, better-integrated and more efficient food grain markets, have reduced seasonal price variations [7].

History of Food Policies

Abolition of Leaky Ration Channels [5]

Food grain distribution under the Public Food Distribution System (PFDS) had expanded rapidly since the liberation of Bangladesh in 1971, reaching its peak in 1989. During the latter part of the 1970s, PFDS distributed almost twice as much food grain as during the latter half of the 1960s. Year to year off-take varied substantially during the 1980s because of natural disasters. The highest distribution was 2.94 million metric tons in the aftermath of the 1988–89 floods.

In December 1991, Government suspended its Palli Rationing channel, the largest in the Public Food Distribution System (PFDS). They formally abolished it, following the leakage rates between 70 per cent and 90 per cent. For a negligible impact on the rural poor, government paid subsidies of 215 crore taka ($60 million) per year. It decided to discontinue the programme.

The annual volume of food grains distribution under PFDS declined from 2.3 million metric tons (mt) in the early 1990s to about 1.4 million mt in the 2000's. Historically, it is believed that the relatively well-off section of the urban population has been the principal beneficiary of subsidised food distributed through the PFDS. The benefits of subsidised food largely bypassed the rural and urban poor, whose need for subsidised food was the greatest. But food grain distribution through better targeting to achieve objectives of poverty alleviation is receiving increasing attention from the government and donor agencies.

Private Import of Food Grains [5]

Since August 1992, government has allowed private import of food grains, initially without any import duty. Private traders responded by bringing in over 300,000 tons of wheat by December of that year. This has allowed a comparable reduction in government's commercial wheat imports. Since government incurs costs of roughly $50 per ton in handling imported wheat, this transfer of commercial imports from public to private agents saved the government about $15 million annually. The shrinking size of government involvement in food grain markets generated considerable cost savings in the food budget, about 254 crore taka ($64 million) annually over the past two years.

With financial savings in hand, and periodically surplus stocks, government has begun experimenting with new, targeted distribution programmes for the poor. The Ministry of Food initiated this effort by commissioning the recent Working Group on Targeted Food interventions to review the effectiveness of all PFDS channels in delivering income support to the poor (Working Group 1994). That review concluded that some PFDS channels - such as the Rural Maintenance Program (RMP) and Vulnerable Group Development (VGD) are very effective in delivering income support to the poor, while others such as the ration channels are not. Drawing on suggestions from the review, Government had launched a large pilot programme Food for Education, in an effort to redeploy savings in ways that will improve the effectiveness of food interventions targeting the poor.

Public issue of all grains as a percentage of availability also shows a negative trend suggesting that the food distribution in Bangladesh has been overwhelmingly privatised following liberalisation. Between 1980 and 2003, the size of the rice markets has grown by about 10 million tonnes.

Three factors, advent of new technology (HYV), development of infrastructure and market liberalisation working in tandem have delivered favourable food security outcomes for Bangladesh. Bangladesh's food-policy has benefited from a liberalised trade regime and a consistent downsizing of the government, all with favourable effects on poverty and nutrition. Post liberalisation, the findings suggest a perceptible increase in the cost-effectiveness of the public food grain distribution system (PFDS). The government has moreover downsized the PFDS, making poverty reduction a priority basis for grain allocation [2].

Expenditure on Food Safety Net

As production increased, the expenditure on food safety net also increased. As shown in table 14, the expenditure has increased from 5.7 per cent of the total budget in 1996-97 to 16.94 per cent by 2008-09. The progress in health indicators show that the increased food production and maybe government subsidies impacted the well-being of some sections of the poor positively. It has yet to reach all but the trickle down effects are noticeable.

Table 13 : Expenditure on food Safety net Programmes as (Per cent of GDP and Budget)

Year	1996-97	2006-07	2007-08	2008-09
Percentage of GDP	0.8	1.4	1.8	2.76
Percentage of Budget	5.7	9.3	13.32	16.94

*(**Source:** Hosaain Sharif (2010), "Does agricultural subsidy in Bangladesh cause agricultural development in Bangladesh?", presented at International Conference Applied Economics, 2010 http://kastoria.teikoz.gr/icoae2/wordpress/wp-content/uploads/articles/2011/10/031.pdf)*

Current Schemes

MDG Fund in Bangladesh (Joint Programmes) [9]

Participating agencies FAO, WFP, UNICEF, Government of Bangladesh

The objective of the joint programme was to contribute to the reduction of underweight prevalence and to reduce the proportion of the population which is food insecure having inadequate calorie and nutrient intake. Elements include homestead food production, in-kind food transfers, community based management of acute malnutrition, school feeding and gardening, and strengthening food security and nutrition information systems. The JPs targeting of agricultural, health and educational inputs in particularly vulnerable coastal areas of southern Bangladesh resulted in small but significant reduction in malnutrition among pregnant and lactating women, and significant improvements in household food security.

The prevalence of severe acute malnutrition (SAM) in children under age five fell to zero from initial prevalence rates of between 2.12 per cent and 0.89 per cent, rising subsequently only to 0.02 per cent and 0.04 per cent. Moderate acute malnutrition fell from between nine and eleven per cent to one per cent with rare occasions of relapse. Malnutrition in pregnant and lactating women using mid-upper arm circumference (MUAC) tapes - an inadequate measure when used alone, fell less sharply–from 7.0 per cent to 5.1 per cent. The programme has also improved household food consumption. Monitoring of the school programme found significant increases in both enrolment and attendance. The government is currently working to update the national nutrition policy and plan of action. Currently, a number of specific policies are place relating to infant and young child feeding, micronutrient provisions and therapeutic treatment for severely malnourished children [9].

Public Food Distribution System in Bangladesh

Originating in 1943 during the Bengal famine, the Public Food Distribution System (PFDS)'s initial objective was to guarantee a minimum quantity of cereals at controlled prices to urban consumers. With a current annual distribution of about 1.5 million metric tons and an average stock level of about 1.0 million metric tons of food grains, the Public Food Distribution System (PFDS) plays an important role in the food and agricultural sectors of Bangladesh. The main objectives of the PFDS are to (1) make food grains available to poor households that would not otherwise have access to adequate food, (2) distribute food during emergency situations, such as natural disasters, (3) provide incentive prices to food grain producers to encourage domestic production, and (4) stabilise market prices to prevent excessive price rises.

The government procures foodgrains from domestic market, imports from abroad, distributes the imported and domestically procured foodgrains through various monetised and non-monetised channels of the Public Food Distribution System (PFDS), sets procurement and distribution prices, strives to maintain floor and ceiling prices in order to stabilise seasonal price fluctuation, and regulates private trade. Over time, food assisted programmes have shifted from relief to development. The development objectives include rural infrastructure development, boosting primary enrolment rates and human capital development. These changes are reflected in the programmes initiated under PFDS, as discussed in the following paragraphs. Almost 80 per cent of the total food grains channelled through the PFDS are now directed towards these programmes compared with 30 per cent in the early 1980s.

As the production increased and the imports markets were privatised, the government could concentrate more on the food subsidies for the poor. About two-third of food assistance programmes are now financed by the Government's own budget as compared to 40 per cent in mid-1990s.

Public distribution essentially involves two kinds of channels, sales and non-monetised. The components of priced distribution are essential priorities (EP), other priorities (OP), open-market sales (OMS) and Large Employers (LE). Priced distribution channel provides subsidised food grains to specific governmental organisation (e.g., defence) or employees of large manufacturing units. Non-priced distribution is motivated by poverty-reduction objectives and includes programmes such as food-for-work programme (FFWP), vulnerable group feeding (VGF), vulnerable group development (VGD), and canal digging (CD), test relief (TR), gratuitous relief (GR), and food for education (FFE). The Food-for-education (FFE) introduced in December 1993 was phased out in December 2001.

However there are still concerns about the efficiency of the PFDS, particularly with respect to the high losses that it incurs at ports, at storage facilities, and during internal transportation [1].

It is important to mention here that the Government of Bangladesh has formulated the National Food Policy Plan of Action (PoA). Apart from providing access to food to the most vulnerable households through targeted cash and food transfers, the Plan also includes actions aimed at developing production capacity, income generation and enhanced nutrition through investments and accompanying policy measures. Thus the programme is wider than the PFDS.

Main Components of PFDS

Food for Work (FFW)

All FFW (Food for work) Programme share the basic objectives of The work developing rural infrastructure and providing employment to the rural poor during the slack season. Beneficiaries are selected through a self-targeting mechanism as individuals can choose whether or not to participate. The programme is theoretically available to everyone, but relatively low wages and labour requirements are set in a way as to discourage the non-poor from participating.

Food-for-Work is not so much a single programme as an umbrella of different programmes and projects carried out by a number of line ministries in collaboration with various donors. Over the years, donor support to the FFW declined somewhat, but the overall size of the programme has continued to rise steadily. For example, in 1999-2000, about 60 per cent of the overall allocation for the FFW was provided by the Government.

The Government runs several FFW programmes without donor support, such as the Rural Infrastructure Reform programme implemented by the Ministry of Disaster Management and Relief. The main donor supported FFW programs are the Rural Development (RD) programme supported by World Food Programme (WFP), bilateral donors, and the Government, and the Integrated Food for Development (IFFD), changed to the cash-for-work (Integrated Food Security Program in 2000) project funded by food assistance from USAID. Food-for-work activities that aim to build community infrastructure are being combined with food-for-training activities that focus on enhancing income-generation capabilities [10].

Food for Education (FFE)

This programme was phased out in 2001. Food- for-Education (FFE) programme was started by the Government on a pilot basis in July 1993. Households with a child in primary school received 15 kilograms of wheat or 12 kilograms of rice a month (those with two or more children got 20 kilograms of wheat or 16 kilograms of rice), provided the children attend at least 85 per cent of their classes and the family sent all children of primary school age to school. The overall allocation to the programme of 285,000 metric tons in 1999-2000, funded entirely from the government's own resources, covered an estimated 2.1 million households.

Mid day meals programmes are replacing this food grain grant programmes in Bangladesh now although the progress of reaching out to the schools is still quite slow [10].

Vulnerable Group Development (VGD)

Started in 1975 as Vulnerable Group Feeding, this programme was renamed as it evolved from providing relief to increasing beneficiaries' self-reliance. The programme, implemented by Government with support from WFP and a number of bilateral donors, comprises three subprojects (i) Union Parishad (UP) VGD (ii) Women's Training Centers (WTC) and (iii) Group Leaders and Extension Workers, of which the first is by far the largest. The projects are exclusively for the benefit of poor disadvantaged women, who receive 30 kilograms of wheat per month and training for 18 months.

The training is different in the different sub projects comprising of: awareness building on social and health issues, functional literacy and numeracy, training in income generating skills, participation in savings schemes and access to credit to start up new income-generating ventures. The WTC trains women for one year in marketable skills on courses operated by the Department of Women's Affairs under the Ministry of Women's and Children's Affairs and several NGOs. The Group Leaders and Extension Workers programme helps organise VGD women in groups, and helps them access extension services (and credit) in areas where the regular package of development services is not available.

There are two steps in the selection of cardholders. The national VGD committee (under the chairperson of the MWCA) distributes cards across the 461 rural thanas. The higher a thana's food insecurity rating by the Government and WFP, the greater the share of cards it gets. A minimum of 30 women are then selected from each union by union and thana level local committees according to selection criteria which include functionally landless women, women with extremely low or no family income, women who are day labourers, and women who lack productive assets. Preference is given to female heads of households (widowed, separated, deserted, divorced or with disabled husbands) and to women who are physically fit to train. Women who receive assistance from other programmes or have participated in the VGD in an earlier cycle cannot be selected for the VGD [10].

Vulnerable Group Feeding and Gratuitous Relief (GR)

Vulnerable Group Feeding (VGF), designed to provide one or more months of food rations to a selected number of households in a period of distress, is implemented by the Directorate of Relief under the Ministry of Disaster Management and Relief. Resources are allocated to the programme by the Ministry of Finance when thought necessary.

Under the programme, food is distributed only among thanas that are affected by disaster. Among the selected thanas, two-thirds of the food is distributed according to population and the remaining one-third is allotted to thanas rated as having very high or high food insecurity on the Government/WFP resource allocation map for food-assisted development. Within thanas, the households to receive food are selected by local communities. Each UP chairman sends a list of potential beneficiaries to a thana-level committee which makes the final selection.

Government selection criteria are defined to include: (i) households in which the head earns less than 300 Tk in a normal month; (ii) households with no agricultural land or productive assets; and (iii) households headed by day-labourers. Priority is given to households that have been physically dislocated by the disaster and to female-headed households. Each household on the list is issued a ration card to receive food for a few months. The Ministry decides how much they receive and for how long, depending on the extent of the disaster. In practice, it is not always possible to restrict food distribution to selected households when other needy households are also present at the UP centre at the time of distribution. In such cases, the distributors sometimes have to reduce the amount given to each cardholder in order to increase the total number of beneficiaries

Gratuitous Relief (GR), designed to provide emergency relief to disaster victims, is also implemented by the Directorate of Relief under the Ministry of Disaster Management and Relief. While GR is a relatively small programme, it is the main way in which the Government provides immediate, short-term relief to disaster areas in form of money assistance [10].

Rural Maintenance Programme

The Rural Maintenance Programme (RMP), while similar to FFW programmes, is smaller than most (50,000 metric tons of wheat equivalent per year) and differs in one important respect: it pays wages in cash, not in kind. It is implemented by local government institutions. Participating women are enrolled in a four-year programme of work and training during which they receive 43 Tk. per day as wages, nearly one-fourth per cent of which is automatically retained in an escrow savings account on their behalf [10].

Open Market sale (OMS) Programme

The Open Market Sale (OMS) is seasonal scheme which provides rice at subsidised price. It normally runs in a year in two different time slots. One slot is identified to be September, October, November and the other slot is identified to be February, March and April. In addition, the programme can be introduced as and when needed. During Ramadan, domestic crisis of food, international market price hike etc., and the government can run the programme as the price hike goes beyond the capacity of the low income consumers. At times of natural calamities and emergency, the programme can be lunched.

The government under its fair price and OMS programmes has sold 23.71 lakh metric tonnes of rice over the last four years (2009-2013) across the country. Each customer can buy a maximum of 5 kg of rice every day. Earlier beneficiaries of this programme were persons below national poverty line. But all are eligible for this programme now since government believes that it will enhance the market competition and reduce the price of rice in the market [27].

Maternal Health Voucher Scheme

Under this scheme, the poor pregnant women receive vouchers that entitle them to free maternal health services, transport subsidies, and a cash incentive for giving birth with the assistance of qualified maternal healthcare provider. It aims at reducing the financial barriers faced by women in accessing health care and increase the demand for institutional services in child birth. It is reported that poor voucher recipients were more likely to deliver in a health facility and two times more likely to use skilled health personnel at delivery than the non-poor recipients. The inequality in accessing health services between the rich and the poor was reported to have decreased even in the short run. However, the authors also report that not all health centres have facilities to provide maternal health care; thus the supply side constraints are hampering the success of the scheme.

Still, the overall impact has been encouraging. The government has initiated expansion of the 'Maternal Health Voucher Scheme' in all upazilas to ensure healthcare services for poor mothers during pregnancy and pre and post delivery period. The 'Maternal Health Voucher Scheme' has already been introduced in 53 upazilas and more 20 upazials would come under the scheme very shortly [8].

The scheme was started in India also in 2005 (it was called Janani suraksha yojana protection of the mother scheme) and was expanded to all states in later years. But from our informal chats with the slum dwellers in Gujarat, we find that not many families are aware of it. The scheme is yet to reach all its beneficiaries.

Assessment of the Impacts of the Above Programmes

Overall participation rates in each of the three programmes (FFE, VGF, VGD) were very low, ranging between three-five per cent of the rural population. Given the low coverage, while the programmes may be extremely important for the beneficiaries themselves, it is unlikely that any one of these could be expected to have a significant impact on overall poverty reduction. While each programme serves only a small part of the poor population, joint coverage rates are higher. Approximately 10 per cent of the population received assistance from at least one of the three programmes in the 12 month period. The limited overlap in coverage by the three programmes conformed with programme targeting criteria for the VGD and FFE programmes that restrict participation to those households that were not already assisted by other food assistance programmes. FFE was slightly better targeted to the poor than the VGD and VGF.

The coverage of the beneficiaries was very limited though, with 85 per cent and 91 per cent of the target populations eligible for the FFE and VGD not included in the programmes. The share of participants who were ineligible for the programme was fairly low for the VGD programme. However, 30 per cent of the FFE beneficiaries did not meet any of the targeting criteria incorporated in the programme design. It is estimated that 71 per cent of FFE beneficiaries reported receiving less than their entitlements, as private dealers responsible for distributing food were diverting grains to the black market. In other instances, there were some reports that dealers were withholding food grains from beneficiaries as payment for prior loans made to beneficiaries at high interest rates. These were some reasons why FFE was discontinued [10].

VGF and GR

While direct distribution through these programmes was quite small, the programmes played an invaluable role by targeting relief to those in need. Immediate short-term relief after the flood was mainly provided by the GR. Only 11.4 per cent of GR recipients were not directly exposed to floods.

However, the amount of transfer provided to beneficiaries was small: on an average households received only 10 Tk per month (over a four-month period), which was much smaller than receipts from private transfers or other government and NGO programmes during the same period. In addition, the programme did not achieve widespread coverage: overall, 69 per cent of flood-affected households did not receive GR transfers. The VGF programme targeted resources to all areas (both flooded and non-flood affected areas) of the country. As a result, while it was not effectively targeted according to flood exposure, it was successful at reaching the poor. Approximately 25 per cent of VGF recipients were not directly exposed to the flood [10].

FFW

In the case of the Food for Work programme, some government departments hired labour contractors who in turn often sold the wheat or rice received so that they can pay cash wages. The workers themselves sometimes end up selling part of the grain they receive to buy other goods and services. As a recent study of the FFE has reported, the dealer based system of food grain distribution was far from satisfactory: households participating in the programme experienced losses in their food grain entitlement due to dealer malpractice, as well as significant transaction costs and inconvenience in collecting their rations [10].

FFE

Previous evaluations of the FFE programme suggest that the numbers of pupils enrolled at participating schools has increased.

The programme increased the probability of going to school by a little over 20 per cent for both boys and girls. The 2000 HIES suggests that gains in enrolment have been sustained. The impact of FFE finds that enrolment and attendance rates are higher in FFE schools, and dropout rates tend to be lower. However, FFE schools have higher student-to-teacher ratios and more crowded classrooms, as a result of which the quality of education is poorer. This point to the need to include complementary financial and technical assistance in order to ensure that student achievement also increases concurrently with higher enrolments [10].

VGD

Evidence on the development impact of the VGD programme is more limited. Comparing the socio-economic status of beneficiaries one year after the joining the programme and two years after leaving the programme. i.e. across a three-year interval, an evaluation had found that beneficiaries of the VGD programme owned more assets (e.g., beds, tin-roofs, clothing), had more savings, and increased access to credit after the programme. Importantly the beneficiaries perceived improvements in their decision-making ability within the family, in the health of their family and in incomes [10].

Concluding Observations on Food Security Schemes

Finding resources to continue food aid in various forms discussed above has been a major challenge for the Government of Bangladesh, particularly since the volume of international food aid received by the country has been declining in recent years. When resources are limited, it is all the more important to ensure that they are used to maximum effect. The food assistance programmes have, by and large, done a good job of reaching the poor. But due to limited resources their reach has remained very small [10].

Still some progress has been made in reducing malnutrition and achieving better health and economic indicators. Interestingly, Bangladesh has been strictly putting away food aid programmes that have large scale leakages. Indian policy makers have not dared to take such drastic steps against "expensive" policies yet. Another feature of the Bangladesh food distribution programme is its heavy dependence on international aid. In this aspect, India has been much more self-reliant. It does not have to worry about stopping the programmes when the outside support stops.

However both nation need to devise better strategies and ways to achieve effective implementation to meet the Millennium development Goals for hunger and malnutrition. Given a very high density of population and wide scale landlessness, they will have to be very innovative in finding solutions.

References

1. Ahmed Akhter, Rashid Shahidur, ,Sharma Manohar, Zohir Sajjad (2004), "Food aid distribution in Bangladesh : Leakage and operational performance", FCND Discussino Paper No. 173, Food Consumption and Nutrition Division, IFPRI, February 2004 http://ageconsearch.umn.edu/bitstream/60392/2/fcndp173.pdf
2. Chaudhry Naimuddin, Roy Devesh and Farid Nasir (2006), "Food policy linearisation in Bangladesh: How markets and government delivered", MTID Discussion Paper No. 92, IFPRI, March 2006 http://ageconsearch.umn.edu/bitstream/58574/2/mtidp92.pdf
3. Dwijen Mallick, Ashraful Amin and Atif Rahman, Climate compatible development in agriculture for food security in Bangladesh, September 2012, https://germanwatch.org/de/download/8347.pdf
4. Indexmundi.com http://www.indexmundi.com/g/g.aspx?v=39&c=in &l=en
5. Haggblade Steven (1994), "Evolving food grain market and food policies in Bangladesh", IFPRI Bangladesh Food Policy Project, May 1994 http://pdf.usaid.gov/pdf_docs/PNABS137.pdf
6. Hossain Mahbub (2013), "Food security in Bangladesh: Achievements and challenges", in The Daily Star, March 20, 2013 http://archive. the dailystar.net/beta2/news/achievement-and-challenges/
7. Hosaain Sharif (2010), "Does agricultural subsidy in Bangladesh cause agricultural development in Bangladesh?", presented at International Conference Applied Economics, 2010 http://kastoria.teikoz.gr/icoae2/wordpress/wp-content/uploads/articles/2011/10/031.pdf

8. One world south Asia (2013), "Maternal Health Voucher Scheme expanded in Bangladesh", June 19, 2013.

   http://southasia.oneworld.net/news/maternal-health-voucher-scheme-expanded-in-bangladesh #.U8WgJPmSyXc

9. MDGFUND.ORG, "Bangladesh", http://www.mdgfund.org/sites/default/files Bangladesh % 20 Joint % 20 Programmes % 20 Fact % 20 Sheet. pdf

10. Murgai Rinku and Zaidi Salman (2004), "Food assistance programme in Bangladesh", Report NO, SASPR2A, SASPR Working Paper, The World Bank http://www-wds.worldbank.org/servlet/WDS Content Server/IW3P/IB/2004/12/06/000012009_20041206112024/ Rendered/PDF/308620 PAPER0SA1tance 0Programs 0in0 BD.pdf

11. Planning Commission of India, "Net Availability of Foodgrain in India" planningcommission.gov.in/sectors/agri_html/selagri/T1.25.xls

12. The news today(2013), "Tk83,987cr spent under social safety net prog" October 29 http://www.newstoday.com.bd/index. php? option = details & news_id = 2360210 & date = 2013-10-29

13. Transform nutrition org, "The hunger and nutrition commitment index 2012 report: Bangladesh situation analysis" http://www.transformnutrition.org/files/2011/11/Bangladesh_situation _analysis. pdf

14. Wikipedia, "Bangladesh" http://en.wikipedia.org/wiki/Bangladesh

15. Wikipedia, "Economy of Bangladesh". http://en.wikipedia.org/wiki/Economy_of_Bangladesh

16. World Bank, "Data: Malnutrition Prevalence, Height for Age", http://data.worldbank.org/indicator/SH.STA.STNT.ZS/countries

17. World Bank (2013), "News: Bangladesh Poverty assessment : A decade of progress 200-2010", , June 20, 2013. http://www. worldbank.org/en/news/feature/2013/06/20/bangladesh-poverty-assessment-a-decade-of-progress-in-reducing-poverty-2000-2010

18. World Bank Social Indicators http://data.worldbank.org/indicator/GC.XPN.TRFT.ZS

19. World Bank Social Indicators

    http://data.worldbank.org/indicator/FP.CPI.TOTL.ZG/countries/BD?display=default

20. World Bank Social Indicators http://data.worldbank.org/indicator/NY.GDP.MKTP.KD.ZG/countries/BD?display=default

21. World bank social indicators http://data.worldbank.org/indicator/SE.ADT.LITR.ZS/countries/BD?display=default

22. World Bank social Indicators http://data.worldbank.org/indicator/SH.DYN.MORT/countries

23. World Bank social Indicators http://data.worldbank.org/indicator/SH.STA.MMRT/countries

24. World Bank Social Indicators http://data.worldbank.org/indicator/SI.POV.NAHC/countries

25. World Bank social Indicatorshttp://data.worldbank.org/indicator/SP.DYN.IMRT.IN/countries

26. World food programme overview, Bangladesh. https://www.wfp.org/countries/bangladesh/overview

27. Zaman Tabassum (2011), "Open market sale- A supply dynamics of two thanas in DHAKA", Department of General and Continuing Education, North South University, Bangladesh. http://mppg-nsu.org/attachments/339_Tabassum%20-Open%20Market%20Sale.pdf

9

Cambodia

Introduction

Cambodia, officially known as the Kingdom of Cambodia and once known as the Khmer Empire, is a country located in the southern portion of the Indochina Peninsula in Southeast Asia. Its total landmass is 181,035 square kilometres (69,898 sq miles), bordered by Thailand to the northwest, Laos to the northeast, Vietnam to the east, and the Gulf of Thailand to the southwest.

With a population of over 14.8 million, Cambodia is the 70th most populous country in the world. The kingdom is a constitutional monarchy with Norodom Sihamoni, a monarch chosen by the Royal Throne Council, as head of state. Cambodia gained independence in 1953. The 1997 coup placed power solely in the hands of Prime Minister Hun Sen and the Cambodian People's Party who remain in power as of 2013. Cambodia is a low income economy with it having one of the lowest annual incomes in the world with the agriculture sector dominating the country's economy, followed by the service and industrial sectors. [9]

Economic and Food Security Indicators

(1) Gross Domestic Product

When we compare the growth rates of Cambodia and India, we find that India has been growing at a faster rate (7.44 per cent on an average) compared to Cambodia (average rate of growth was 5.85 per cent). The standard deviation of the growth series is also lower for India (1.74) compared to Cambodia (3.56). Thus Indian economic growth pattern is more stable than Cambodia and higher. India is thus doing better than Cambodia.

Table1: GDPGrowth Rate of India and Cambodia

Country	1999	2000	2002	2003	2004	2005	2006	2007	2008	2009	2010	2011
India	5.5	6	4.3	8.3	6.2	8.4	9.2	9	7.4	7.4	10.4	7.2
Cambodia	4	4	5.3	5.2	5.4	13.4	7.2	10.1	5	-1.5	6	6.1

Source: *CIAWorldFactbook,2011 (http://www.indexmundi.com/g/g.aspx v= 66 & c=cb&l=en)*

India: MEAN- 7.44 STDEV= 1.738577

Cambodia: MEAN- 5.85 STDEV=3.561026

(2) Inflation Rate

The rate of inflation has been higher in India (6.76 in last 12 years) compared to Cambodia (4.95). However, the standard deviation for Cambodia is much higher (5.04) compared to India (2.63). This shows that although the average rate of inflation seems lower for Cambodia, the stability in prices and lack of volatility is actually higher in India.

Table 2: Inflation Rate of India and Cambodia

Country	1999	2000	2002	2003	2004	2005	2006	2007	2008	2009	2010	2011
India	6.7	5.4	5.4	3.8	4.2	4.2	5.3	6.4	8.3	10.9	11.7	8.9
Cambodia	4.5	1.6	3.3	1.7	3.1	5.8	5	5.9	19.7	-0.7	4.1	5.5

Source: *CIAWorldFactbook,2011(http://www.indexmundi.com/g/g.aspx?v=71&c=in&l=en)*

India- MEAN- 6.766 STDEV=2.634158-India

Cambodia-MEAN- 4.95833 STDEV=5.047674

3) Infant Mortality Rate

IMR is lower for India (46.07) compared to Cambodia (54.08). The rate of fall (27 per cent over last 11 years) in IMR was also faster for India, compared to Cambodia (17 per cent). The paired t test for the series available for IMR for both the nations showed a highly significant difference between the two time series (the t statistics was 5.87 and p was highly significant at 0.0001.

Table 3: Infant Mortality Rate of India and Cambodia

Country	2001	2012
India	63.19	46.07
Cambodia	65.41	54.08

Source: *CIAWorldFactbook,2011(http://www.indexmundi.com/g/g .aspx? v=29& c=in & l=en)*

Source:CIAWorldFactbook, 2011 (http://www.indexmundi.com/g/g. aspx ? v = 29 & c = cb & l = en)

(4) Maternal Mortality Rate (MMR)

Cambodia has managed to reduce its MMR and hunger levels at much faster rate than India, as the table 4 shows. The MMR was more than double in Cambodia in 1990. From 1200 MMR, when (India's MMR was 560), Cambodia reduced it to just 170. With that, Cambodia has twenty points lower MMR than India (190). India needs to learn about the schemes used by Cambodia to address its problem of maternal mortality.

Table 4: Maternal Mortality Rate of India and Cambodia

Year	1990	1995	2000	2005	2013
India	560	460	370	280	190
Cambodia	1200	860	540	320	170

Source: *UNICEF ChildInfo Monitoring the Situation of Children and Women, 2014(http://www.who.int/gho/maternal_health/countries/ind.pdf?ua=1)*

(http://www.who.int/gho/maternal_health/countries/khm.pdf?ua=1)

Poverty and HDI

The poverty levels also have fallen at a slightly higher rate for Cambodia (table 5.1 and 5.2). The proportion of people living below the poverty line reduced from 47 per cent to 30 per cent between 1995 and 2008. The rates are similar and the reduction is almost the same in India: the number declined from 45.3 in 1994 to 29.8 in 2010.

Thus both are at almost the same level. The values of human development index are also comparable: 0.554 for India and 0.543 for Cambodia. Both started at around 0.410 in early 1990s. Thus overall performance of both the nations is similar, as the HDI scores (tables Six and Seven) show. The life expectancy is similar at 65 years for India and 63 for Cambodia. However, the estimates of life expectancy in India by other agencies has been 72, but for the sake of comparison, we can keep these figures.

In reducing undernourishment and child malnutrition, Cambodia achieved notable success over the past two decades. Despite setbacks during the Asian crisis, the prevalence of undernourishment was brought down from 38 per cent in 1992 to 25 per cent in 2006. While 59 per cent of all children under five years of age were stunted and 43 per cent underweight in 1996, the indicators declined to 40 per cent for stunting and 29 per cent for underweight in 2008. India's record has been much poorer for under nutrition, as our data from the chapter on Bangladesh shows: 42 per cent were under weight in 2010 and 43 per cent were stunted.

Although progress in reducing child mortality has gathered pace, particularly since 2000, Cambodia seems unlikely to meet the fourth Millennium Development Goal (MDG) of reducing the under-five mortality rate by two thirds between 1990 and 2015. In 1980, 150 out of 1,000 children younger than five years of age died from malnutrition, poor health, or other preventable causes in Cambodia. The mortality rate declined to 117 in 1990 and 90 in 2008. While this progress is impressive, the gap between the achievement and the target of the fourth MDG is significant. [5]

India had under five mortality rate of 61 in 2010, as our data from Bangladesh chapter shows. Thus the stunting and mortality rates are similar to Cambodia, but the underweight ratio is much higher for India.

Table 5.1 : National Poverty Rates of Cambodia.

Year	1995	1998	2005	2008
Cambodia	47	36.1	34.7	30.1

Table 5.2: National Poverty Rates of India.

Year	1994	2005	2010
India	45.3	37.2	29.8

Source: *Food Security Portal, 2012 (http://www.foodsecurityportal.org/api/countries/per cent-below-povert)*

Table 6: HDI Values for India

Year	Life Expectancy at Birth	Expected Years of Schooling	GNI per capita (2005 PPP$)	HDI value
1990	58.3	7.4	1,191	0.410
1995	59.8	8.2	1,389	0.438
2000	61.6	8.3	1,702	0.463
2005	63.3	9.9	2,190	0.507
2010	65.1	10.7	3,009	0.547
2011	65.4	10.7	3,175	0.551
2012	65.8	10.7	3,285	0.554

Source: *Human Development Report, 2013(http://hdr.undp.org/sites/default/files/Country-Profiles/IND.pdf)*

Table 7: HDI Value of Cambodia

Year	Life Expectancy at Birth	Mean Years of Schooling	GNI per capita (2005 PPP$)	HDI value
1995	56.2	5.5	0979	0.411
2000	57.6	5.7	1002	0.444
2005	60.1	5.7	1440	0.501
2010	62.7	5.8	1868	0.532
2011	63.1	5.8	1988	0.538
2012	63.6	5.8	2095	0.543

Source: *Human Development Report, 2013(http://hdr.undp.org/sites/default/files/Country-Profiles/KHM.pdf)*

History

Consideration of current food security circumstances in Cambodia needs to be put into the context of the country's socio-economic and political development. In terms of economic development the country has been, and still is, underdeveloped in comparison with many Asian countries. The manufacturing, mineral extractive and infrastructural base in Cambodia has always been slight. It began to develop in the early 1960s, but this economic development was cut short by political developments. The economy has always been primarily based on agricultural production. Although agricultural production has, in times past, expanded on the basis of irrigation and drainage systems, it has only marginally been affected by the technological advances in agriculture that other countries have experienced during the last 25 years.

About two thirds of this area is used for rice cultivation. The population (about 13.4 million in 2008) is limited in relation to cultivated land area. The production system has been predominantly one of smallholders, without developed agricultural extension, credit, input and output services. This has mainly resulted in extensive rather than intensive farming systems. Nevertheless, until the early 1970s, the agricultural sector managed to make available sufficient food to cater for the basic needs of much of the population, and also to provide food for export during the 1960s.

In the late 1960s, Cambodia was drawn into the "Cold War" Sino-Soviet and Sino Vietnamese conflicts. It had disastrous consequences for the country in terms of human life and social and economic development. It was largely these conflicts, continued during the 1980s by the forces opposed to the Soviet-Vietnamese alliance, which caused such hardship in Cambodia right up until the signing of the Paris Peace Accords in 1991. The social disruption, economic dismantling and decay, and human loss of life caused by the Democratic Kampuchea (DK) ("Khmer Rouge") Government (1975–1979) were substantial, and the impact of this period on current life is still being strongly felt.

During 1970s, the forced migration of people to and within rural areas, and a focus on food production by the resettled population, was in theory aimed at improving food security. Initially rice production increased, but the elimination of the limited number of people with technical skills, poorly undertaken irrigation schemes and continued extraction of food for exports led to a sharp reduction in food supply. The difficulties faced by subsequent governments in recovering from that period have been considerable.

Till 1990, the war continued at considerable human and financial cost. The development style during this period was very much a government led, government-controlled approach, with limited scope for private sector actors or social structures run or developed by organisations other than the state. In terms of food security, however, a basic food supply to urban areas was largely ensured through a procurement and rationing system. Land redistribution and collective structures enabled many rural households to supply their basic food needs, particularly vulnerable ones such as widow-headed households with many dependants, of which there were many during this period. Towards the end of the 1980s, despite all these difficulties, Cambodia appeared to be approaching national food self-sufficiency again.

Cambodia has experienced rapid economic growth since 1993, especially after the turn of the millennium. Between 1993 and 1998, the GDP grew by 6.3 per cent annually and by 9.5 per cent between 1999 and 2008. For the agricultural sector, despite negative growth in 2000 and 2002, the sector's value added grew at 4.6 per cent between 1993 and 2008 with little differences before and after the Asian crisis. Even more importantly, as suggested by FAO's food production index, food production increased by more than 2.5 times since 1993 (FSS 2010). However, Cambodia's population is also one of the fastest-growing in South and Southeast Asia. From 1993 to 2008, population grew at 2.3 per cent annually. The number of people who are economically active in agriculture grew more rapidly, with an annual rate of 3.0 per cent. Given high population growth rate in the rural area, the agricultural value added per agricultural worker grew by 1.9 per cent only, in contrast to the impressive per capita GDP growth rate of 6.3 per cent annually. Income inequality, measured by the GINI coefficient, has been worsening in Cambodia during its rapid growth period. In 1994, the GINI coefficient was

38.3, and in 2007, it was 44.2. Thus, Cambodia becomes a country with the most rapid increase in GINI coefficient in South and Southeast Asia today. [8]

In spite of the rapid increase in income inequality, poverty declined tremendously, from 47 per cent in 1994 to 30 per cent in 2007. While progress has been made in both rural and urban areas, the poverty rate is still as high as 35 per cent in the rural area in 2007, much higher than the urban poverty rate. Fast economic growth in urban areas and relatively slow progress in rural poverty reduction might have contributed to increasing countrywide income inequality. [6]

Policy Over Years

The radical governance of the Khmer Rouge from 1975 to 1979 led to widespread malnutrition and poverty. It was not until the formation of the Royal Government of Cambodia in 1993 that new policy frameworks could be progressively developed. These included the 1994 National Programme to rehabilitate and Develop Cambodia (NPRD), the first five-year Socio-Economic Development Plan (SEDP I, 1996–2000), the SEDP II for the period 2001–2005 and the National Poverty Reduction Strategy (NPRS) for the period 2003–2005.

The SEDP II (2001–2005) and the NPRS (2003–2005) specifically addressed the crosscutting issue of Food Security and Nutrition (FSN) and its mainstreaming in national and sector strategies. Projects focusing on FSN policy support, capacity development, improvement of information management and creation/strengthening of coordination structures were developed. In 2005 the Technical Working Group (TWG) on FSN chaired by CARD participated in the NSDP 2006–2010 consultation process to provide inputs with regard to the cross-cutting issue of FSN. A strategy paper which forms the basis of this SFFSN was developed in 2006.

The SFFSN is expected to provide substantial progress toward improved food security and nutrition conditions in Cambodia. Based on this strategy, by 2020, the poor and food insecure Cambodians should have substantially improved their physical and economic access and are expected to have sufficient, safe and nutritious food to meet their dietary needs and food preferences for an active and healthy life. [6]

While these policy frameworks were an achievement, they frequently suffered from insufficient coordination and institutional capacity for implementation. In terms of FSN (Food Security and Nutrition), their impact was limited by insufficient specification of the nature and prevalence of food insecurity and malnutrition and their causes, and the lack of a comprehensive strategy to address these problems. It should be recognised that while these national policy frameworks and relevant sector policies were being developed, a number of programmes aiming to increase food security were already under way, with some of them dating from the 1980s. [1]

Direct Food Subsidies or Distribution

Even before the 2007–08 food crisis, the government had begun considering various policies and programmes to increase agricultural production through productivity enhancement and increased value addition especially in the rice production, marketing and export chain. In Cambodia, the government does not provide direct price support to farmers and intervenes only marginally in the market. However, VAT exemptions on agricultural inputs are available and some recent decisions have addressed the necessity to ensure access to seeds and fertilisers at affordable prices, especially in times of emergency. The Law on Farmer Cooperatives was promulgated in 2013. It defines the establishment, functioning and management of farmer cooperatives and provides a framework for better coordination. [2]

Distribution of food and provision of food subsidies have been undertaken frequently but only in emergency context, especially under the Emergency Food Assistance Project (EFAP). EFAP is a major operation that took place in October 2008 to mitigate the effects of soaring food prices on poor households with in-kind food assistance disbursed in eight provinces surrounding the Tonle Sap Lake. Also during the massive floods in 2011, the government responded by distributing food to the worst affected provinces and rice seeds to farmers. [2]

[http://www.fao.org/docrep/field/009/i3761e/i3761e.pdf]

Rural development and improvement of the livelihood of the rural poor have been recognised as top priorities of the NSDP 2006–2010. In particular, social protection is one of its key commitments, including creating jobs in the formal and informal sectors and ensuring improved working conditions; providing social safety nets for the disadvantaged; measures to assist victims of natural disasters; targeted programmes for vulnerable groups such as poor female-headed households and veterans, and establishment of rehabilitation centers for orphans and the elderly; addressing domestic violence and trafficking; and improving health service delivery, quality, and financing. In addition, the NSDP explicitly recognises the need for further progress in food security and nutrition as a key action to meet the NSDP priorities.

Policies

(1) SFFSN (Strategic Framework for Food Security and Nutrition)

The Strategic Framework for Food Security and Nutrition in Cambodia 2008-2012 (SFFSN) is a reference document on the cross-cutting issue of food security and Nutrition. Improving food security and nutrition (FSN) is a development priority of the Royal Government of Cambodia (RGC), as is apparent from its current national strategic frameworks, including the Cambodia Millennium Goals (CMDGs), the Rectangular Strategy and the National Strategic Development Plan (NSDP) 2006-2010. It is also a priority recognised in a broad range of sector policies, from agriculture and water resources to economic development and health and education sectors.

In 2004, 34.7 per cent of Cambodians were below the poverty line and 90 per cent of these poor were located in rural areas. Nationally, 23 per cent of the Cambodian population, or three million people, were food-deprived in 2003–2004, consuming less than the minimum daily energy requirement of 1715 kcal/day. In 2005, more than 630,000 or 37 per cent of Cambodian children aged under five were suffering chronic malnutrition (stunting) and seven per cent of children were acutely malnourished (wasted). More than 60 per cent of children under two suffered from anaemia. Malnutrition is also a major cause of the high level of maternal and infant mortality. [1]

Objective 1: Increase Food Availability

SFFSN strives to increase food availability by making food-insecure households increase food availability from their own agriculture and livestock production and from common property forests and fisheries. This is because most poor and food-insecure households in Cambodia are rural small hold farming households. Traditionally, these households have used the above mentioned sources for securing food supply. However, rural Cambodians typically confront low productivity and high risks in their own efforts in food production and harvesting forests and fisheries. The following activities are planned for achieving this objective:

- Increase rice productivity, crop diversification, and productivity of non-rice crops, upland agriculture systems and livestock.
- Improve available water supply for crop and livestock production through expanding irrigation and agricultural water supply
- Allocate demined land to the rural poor and food-insecure.
- Expand low-cost, environmentally friendly rice intensification (SRI) for poor food-insecure farmers.
- Targeting food-insecure areas (low land plains and uplands what suffer the most from droughts and floods) and social groups (small hold farmers and women) in irrigation investments.

Objective 2 Increased Food Access

SFFSN would try to help the food-insecure households to increase their food access by increasing household income. The rural poor and food-insecure are a numerically large component of the private sector–specifically in the informal sector. Improving household incomes and employment for this group would greatly enhance food security situation. The following actions are planned for achieving this objective:

- Urban government authorities to scale up vocational and micro-enterprise training for the urban poor who are dependent on casual wage labour for income.
- MoC (Minister of Commerce) to establish a strategy for informal-sector household micro-enterprise development.

- Ministry of Agriculture, fisheries and forests (MAFF) together with Ministry of Information and the mass media to increase the regular flow of information on agricultural, livestock, forest and fisheries products and labour markets to small hold farmers by radio.

Objective 3: Improvement in Health and Nutrition Status

This SFFSN Objective focuses primarily on improvement of the health and nutrition status of women and children. The following actions are planned:

- Increase the rate of early initiation of breastfeeding, including colostrums, within one hour of delivery.
- Increase the rate of exclusive breastfeeding of infants in the first six months.
- Provide de-worming for preschool and school children.
- Promote integrated community based nutrition programmes to improve nutritional knowledge and practices.
- Provide micronutrient supplements (iron) and fortified food to primary school children.
- Reduce the low birth weight rates by improving maternal nutrition.
- Continue implementation of the MoH Health Sector Strategic Plan 2003-2007 to promote higher quality and more affordable health care services for the rural poor, to provide technical training for health care providers and to strengthen monitoring and regulation of public and private sector health care services.

Objective 4: Increased Stability of Food Supply

The aim is to improve social safety nets and enhanced capacities of food insecure households to cope with natural disasters as well as unforeseen crisis. And thus increase stability of their food supply. The activities planned are:

- Expand the disaster preparedness, mitigation, response and recovery activities of Red Cross agencies, other NGOs, and government/donor collaborative rural development projects which are already established and operating in the country.

- Increase training/extension/inputs to vulnerable communities on better flood, drought, pest and disease-resistant crops, through increased MAFF/Disaster Management agency collaboration.
- Continue and scale up targeted social transfer systems in cooperation with international development agencies, NGOs and donors.

5. Enhanced Institutional and Policy Environment for FSN

It aims at enhancing the ability of the state to address and overcome food insecurity and malnutrition. The coordination among stakeholders is also sought to be improved. The activities to be undertaken for achieving this objective are as follows:

- Build capacity of the TWG-FSN Secretariat within the Council for Agricultural and Rural Development (CARD) to facilitate inter-ministerial networking and coordination on FSN issues.
- Develop capacity, procedures and mechanisms to more comprehensively incorporate FSN issues into the local government planning process.
- Investigate methods and approaches for increasing the frequency and specificity of FSN primary data collection and increasing knowledge of variations in FSN outcomes and their causes by geographic area and socio-economic group within Cambodia.
- Provide regular updates of the Cambodian Food Security Atlas (WFP) and promote its use for policy analysis and area targeting.[1]

2) NSDP (The National Strategic Development Plan)

Ministry of Planning (MoP) has the task to prepare a National Poverty Reduction Strategy in order to qualify for Poverty Reduction Credits from International Monetary Fund (IMF) and World Bank (WB). In coordination with line Ministries and External Development Partner (EDP) the MoP therefore prepared the National Poverty Reduction Strategy Papers which resulted into National Strategic Development Plan (NSDP) 2006-2010. The NSDP was developed by the MoP with participation and consultation from all stakeholders including government Ministries and institutions, External Development

Partners (EDPs) and civil society. It is intended to align sector strategies and planning cycles to overall long term vision, as well as guide external development partners (EDPs) to align and harmonise their efforts towards better aid-effectiveness and higher 'net resources' transfer than hitherto. After approval by the National Assembly, the Senate, the RGC and the King the NSDP was officially launched on 15th of August 2006 by the Prime Minister. [3]

Major goals mentioned in NSDP are:

- Eradicate poverty and hunger,
- Develop the agriculture sector and enhance agricultural production and productivity,
- Implement the Education Sector Strategic Plan,
- Implement the Health Sector Strategic Plan,
- Implement population policies.
- Further advance rural development,
- Ensure environmental sustainability,
- Promote gender equity,
- Implement good governance reforms,
- Sustain high macroeconomic growth,
- Improve budget performance,
- Accelerate industrial growth,
- Further develop the private sector,
- Increase trade (i.e., export),
- Develop tourism,
- Make progress in de-mining and provide victim assistance,
- Rehabilitate the physical infrastructure,
- Further develop the energy sector.

Based on projected macro-economic growth and needed investments in the public sector, and taking into account the absorptive capacity of the public sector apparatus, it is estimated that Cambodian Riel (CR) 15,225 billion equivalent to US$ 3,500 million will be needed for the public sector during 2006-2010 to achieve the goals and targets set forth in the NSDP.[4]

Conclusion

The achievement in reducing malnutrition and poverty in Cambodia seems to have come more from (a) the country returning to conditions of a peaceful economy and (2) the resulting rapid economic growth due to such conducive environment. There seems to be little investment on malnutrition schemes, at least till early 2000s. In this sense, the path of growth has been working in reducing malnutrition till now. To reduce the malnutrition levels further, the country needed more converted efforts, which seem to have been initiated since 2006 in the form of SFFSN and NSDP. However, none of these plans are as ambitious as India's Food Security Bill. We need to see how Cambodia progresses in achieving lower levels of malnutrition and stunting with much smaller investments in food security programmes. If the progress is found to be comparable to India, then there are lessons to draw in terms of the size of interventions needed for achieving food security.

References

1. Council for Agricultural and Rural Development, "Strategic Framework for Food Security and Nutrition in Cambodia 2008-2012 (SFFSN)",May 2008(http://www.gafspfund.org/sites/gafspfund.org/files/Documents/Cambodia_4_of_16_STRATEGY_Food_Security_and_Nutrition_Strategy_0.pdf)

2. Food and Agriculture Policy Decision Analysis, "Country Fact Sheet on Food and Agricultural Policy Trends", 2014 (http://www.fao. org/docrep/field/009/i3761e/i3761e.pdf)

3. Ministry of Planning, "National Strategic Development Plan (NSDP)", 2006 (http://www.mop.gov.kh/Home/NSDP/tabid/83/Default.aspx)

4. Ministry of Planning, "National Strategic Development Plan (NSDP)- report 2006-2010", 2006 (http://www.mop.gov.kh/Home/NSDP/tabid/83/Default.aspx)

5. Olivier Ecker, Ph.D. Post-doctoral Fellow, Xinshen Diao, Ph.D. Senior Research Fellow, "Food Security and Nutrition in Cambodia: Patterns and Pathways", 2011 (http://www.cdri.org.kh/webdata/download/sr/foodSecuNutrition10e.pdf)

6. Olivier Ecker, Ph.D. Post-doctoral Fellow, Xinshen Diao, Ph.D. Senior Research Fellow, "Food Security and Nutrition in Cambodia: Patterns and Pathways", 2011(http://www.cdri.org.kh/webdata/download/sr/foodSecuNutrition10e.pdf)

7. Paolo Santacroce, World Food Programme, VAM Food Security Analysis, "Kingdom of Cambodia: Comprehensive Food Security and Vulnerability Analysis", Dec 2008 (http://home.wfp.org/stellent/groups/public/documents/ena/wfp227417.pdf)

8. Vincent Tickner, "Food Security in Cambodia- A Preliminary Assessment", 1996 (http://www.unrisd.org/80256B3C005BCCF9/(httpAuxPages)/64B8DCC549D8524980256B67005B6C36/$file/dp80.pdf)

9. Wikipedia, "Cambodia", 2014 (http://en.wikipedia.org/wiki/Cambodia)

10

Pakistan

Compilation: Nirav Devpura

Introduction

Pakistan is a low income country where agriculture plays an important role in providing employment and food to its large population, which is touching 180 million now, and increasing with a growth rate of 2.09 per cent. According to this growth rate, Pakistan could become 4th most populous country in 2050 (Government of Pakistan, 2010). Along with the population growth, land under cultivation has also increased to 40 per cent since 1960. Agriculture provides employment to the two-third of the population contributes 22 per cent in the GDP and constitute 60 per cent of the total export in Pakistan. About 68 per cent people in Pakistan live in the rural areas. Although, Pakistan enjoys self- sufficiency once in a while, according to Planning Commission of Pakistan, number of people below poverty line was about 64 million in 2008. It is because the food security policies and initiated by Government of Pakistan have not been implemented appropriately.

Economic Indicators

The per capita income of India of around $ 1500 was somewhat higher than Pakistan's $1200 in last three years (Table 1).

The rate of growth of per capita income for India during the last 10 years has been calculated by the semi log regression:

Log (Per Capita Income) = $\alpha + \beta$ (Year) where β directly gives the growth rate

The rate of growth of per capita income was 10 per cent in India, according to the semi-log regression (R square = 0.92 and t statistics is significant at 10.24). A similar analysis for Pakistan shows the growth rate for per capita income to be 7.68 per cent.

The results of the paired t-test show that the per capita income of India is statistically significantly higher than Pakistan (t statistics = 3.46, p value = 0.003).

The GDP growth rates are also quoted for a few years (Table 2) to show that the Indian growth rates are consistently much higher than those of Pakistan.

Table 1 : GDP Per Capita Income

Year	India	Pakistan
2004	650	631
2005	740	693
2006	830	853
2007	1069	930
2008	1042	1018
2009	1147	987
2010	1417	1023
2011	1540	1213
2012	1503	1255
2013	1499	1299

*(**Source** : http://data.worldbank.org/indicator/NY.GDP.PCAP.CD)*

Table 2 : GDP Growth Rates (Per cent)

	Pakistan	**India**
2005	7.5%	9.3%
2007	4.8%	9.8%
2009	2.8%	8.5%
2011	2.8%	6.6%
2012	4%	4.7%

*(**Source :** http://data.worldbank.org/indicator/NY.GDP.MKTP.KD.ZG)*

Social Indicators

The social indicators also are higher in India - in 2011, India had literacy rate of 73 while Pakistan's was 57 (Table 3). The life expectancy also is slightly higher for India (70) as opposed to Pakistan's 66 (Table 4).

Table 3 : Literacy Rate Above 15 Years of Age.

	Pakistan	India
2001		64.8
2005	50	
2006	53	62
2008	54.4	
2011	57	73

(***Source :*** *http://data.worldbank.org/indicator/SE.ADT.LITR.ZS)*

Health Indicators

Given the background of economic and social indicators, one would expect the health indicators to be lower for Pakistan compared to India; we found that indeed is the case. The mortality rate for children under 5 years of age was 86 for Pakistan in 2012 whereas it was 56 for India (Table 5.1). The paired t test shows the difference in the series to be highly significant with t statistics to be as high as 49.4. Similar is the case for infant mortality rate (t statistics = 83.5) given in Table 5.2: India has significantly better IMR (44) compared to Pakistan (69). As shown in Table 4, the life expectancy is also higher in India (70) compared to Pakistan (66).

Table 4: Life Expectancy at Birth, Total (Years)

	Pakistan	India
2004	65	64
2005	65	64
2006	65	64
2007	66	65
2008	66	65
2009	66	65
2010	66	67
2011	66	68
2012	66	70

(***Source :*** *http://www.indexmundi.com/facts/india/life-expectancy-at-birth)*

Table 5.1: Mortality Rate Under 5 (per 1,000 live Births)

	Pakistan	India
2004	103	78
2005	101	75
2006	98	72
2007	96	69
2008	94	67
2009	92	64
2010	90	61
2011	88	59
2012	86	56

*(**Source :** http://data.worldbank.org/indicator/SH.DYN.MORT)*

Table 5.2: Mortality Rate infant (per 1,000 live births)

	Pakistan	India
2004	81	58
2005	80	56
2006	78	54
2007	77	52
2008	75	51
2009	74	49
2010	72	47
2011	71	45
2012	69	44

Malnutrition Indicators

But in spite of all this, the figures in Table 6 for malnutrition are lower for Pakistan (31, compared to 42 of India). The data points are not enough for running a statistical test. But the numbers show that the rates are lower for Pakistan both for percentage of malnourished children under the age of five (Table 6) as well as according to the Global Hunger Index (Table 7).

The per capita food availability is also somewhat higher in India at 165 as opposed to 153 in Pakistan (Farooq, et al, 2009). It is difficult to understand how more children die in Pakistan than India, how general life expectancy is higher in India and yet children can be better nourished in Pakistan compared to India. Somewhere some statistics seem to have gone wrong to create this contradiction.

Table 6: Malnutrition Prevalence, Weight for Age (Per cent of Children Under 5)

	Pakistan	India
1979	48.3	67.3
1990	42.4	59.5
1992	44.5	50.7
1997	33.2	41.1
1999	32.8	44.4
2006	31.1	43.5
2011	30.90	42.3

*(**Source :** http://www.indexmundi.com/facts/pakistan/malnutrition-prevalence)*

Table 7: Global Hunger Index

	India	Pakistan
1990	31	25.9
1996	22.4	22.8
2001	23.1	21.2
2012	22.1	19.3

*(**Source:** http://en.wikipedia.org/wiki/Global_Hunger_Index)*

Agriculture in Pakistan

Pakistan has made significant progress in terms of increasing food supplies. Per capita availability of cereals increased from 120 kilograms in 1961 to 137 kilograms 1990-91 and further increased to 154 kilograms in 2008-9 of which, more than 80 per cent is accounted for by wheat alone.

In Pakistan, major crops such as wheat, rice, cotton and sugar cane account for 82 per cent of value added in the crop sector, 33 per cent of value added in overall agriculture and seven per cent of GDP. Wheat alone accounts for 14 per cent of value added in agriculture and provides three per cent of the country's GDP according to the Ministry of Finance (2009). Wheat is the most important agricultural crop; it is grown by about 80 per cent of farmers on about 9 million ha, which is close to 40 per cent of the country's total cultivated land. The crop also accounts for an estimated 37 per cent of both food energy and protein intakes. It is therefore the single most important food crop in Pakistan.

Pakistan government procures wheat from farmers to guarantee profitability to the wheat farmers. The procured wheat is sold to the millers at a rate that would ensure stable prices for flour and bread.

There has been strong government involvement in the wheat sector since the late 1950s. In 1959/60, the government fixed the domestic prices for wheat and other crops at higher than international prices to promote domestic production. During this period, a ban was placed on interprovincial wheat trading by the private sector, to facilitate government procurement of wheat from surplus production areas. Pakistan has maintained a heavily controlled wheat sector since then. A portion of the crop is purchased by PASSCO and provincial food departments, at above market price, selling it afterwards to flour mills at lower price. By purchasing at higher price, the government offers an incentive to growers, while selling it to millers at lower rate ensures cheap availability of flour and bread in urban areas.

The federal government establishes the procurement price at the beginning of each marketing year, together with targets for wheat procurement by provincial food departments and federal agencies. The government also sets the release price, which can be adjusted later during the marketing year.

Every new government experiments with the wheat policy, at the expense of the people, and there is no consistency at all. At times there is surplus wheat which rots away in open storages. At other times the supply is so short that the government imposes all kinds of unenforceable restrictions which only result in black marketing, exploitation of the farmers, and high prices for the customers.

The system of wheat subsidy is manipulated by PASCO staff, land revenue officials, and middlemen, to the detriment of government, taxpayers and farmers. According to the President of the Farmers Association of Pakistan, Tariq Bucha, this year, farmers have lost 10 to 20 billion owing to the government's poor purchase policy and unfair gunny bags distribution.

Pakistan is not consistently self- sufficient in food. It needs to import wheat often. In such a situation, the consumers' interests are hurt when the issue prices are above the import parity price. This becomes even more true considering that only about one-fifth of farming households produce surplus wheat, and many farmers have to buy wheat as they do not produce enough to meet their entire year's demand. [12], [13], [17], [18], [19]

While the people do not benefit, the financial losses suffered by the government are also considerable. To take an example of the financial year 2009-10.

- Support Price : 950/40 (23.75 Rs/kg)
- Official Maintenance costs : 196/40 (4.9 Rs/kg)
- Release Price : 975/40 (24.38 Rs/kg)
- Subsidy on wheat procured : 171/40 (4.28 Rs/kg)
- Potential Losses at 2009-10 release price : 39.5 billion Rupees

Chronology of Major Events in Pakistan Wheat Policy and Markets [1]

Independence to early 1980s: substantial government market intervention ration shops with fixed prices, but substantial leakages and malpractices.

- November 1987: abolition of wheat ration shops.
- Late 1980s: broad trade liberalisation; private sector wheat imports allowed in late 80s, but subsequently disallowed.
- Bumper harvest in 1999-00 (i.e., March-April 2000)
- 8.6 million tonnes procurement; large increase in stocks.
- Subsequent subsidised exports of stocks, including private sector exports.
- Incentives for private investment in storage.
- Public investments in laboratories for grain testing.
- Crop shortfalls in 2001-02, 2002-03 and 2003-04.
- Government policy response to shortfalls: 2004:

- Restrictions on inter-provincial transport of wheat (and inter-district transport of wheat in Punjab).
- Government tenders for imports (but some shipments rejected on quality basis in early 2004).
- Increased procurement price for 2004-05 crop (to 400 Rs/40 kg).
- Production recovery and market liberalisation: 2005
- Good harvest 2004-05 (21.1 million tons).
- Lifting of restrictions on transport of wheat.
- Encouragement of private sector commercial wheat imports.

Policy Scenario, Devolution Process and Institutionalisation of food Security

Pakistan is a federation of four provinces and federally administered areas. The style of governance in the country has been predominantly federal, in which many sectors were directly under the federal ministries. Provinces had limited decision powers. However in 2011, the elected parliament approved a land mark 18th constitutional amendment, by which powers for certain matters in concurrent list were transferred from centre to the provinces. Since June 30, 2011 seventeen ministries and divisions have been devolved and their authorities have been transferred to the provincial governments. Food and Agriculture was one of it that was transferred.

According to the experts, the devolution of the formal federal Ministry of Food, Agriculture and Livestock (MINFAL) would create serious issues relating to food security, international co-operation and agriculture research as the provinces with the exception of Punjab do not have required infrastructure and capacities to deal with the food security operations. This raised concerns all around. Many argued that agriculture is a federal subject around the world and therefore the government should not completely decentralise the ministry. Following these concerns a new federal ministry with the title "National Ministry of Food Security and Research (NMDSR), was created in October 26, 2011, with a specific mandate of protecting food security of the people of Pakistan and collaborate with international actors.

A public sector development programme with a financial outlay of Rs. 495 million has been initiated for the projects of newly created ministry of National Food Security and Research. These projects reflect the above referred official policy elements, such as research and development, emergency food delivery, seed improvement, infrastructure development and grain storage. [8]

Food Security Schemes and Programmes

Ration Shops

Pakistan's ration system was established in 1942 to deal with shortages of basic goods caused by wartime disruption in supply. At that time, the ration shops handled wheat and sugar, tea, matches, kerosene, yarn, and cotton cloth. After partition, the system was continued to control hoarding and profiteering of scarce goods. All trade in wheat was rationed and controlled by the government until the 1960s when, as a result of several years of favourable weather, supplies became plentiful. Rationing was abolished, but the shops continued to sell atta (whole wheat flour), which the government obtained through its guaranteed price support scheme, with no limitation on quantity.

In the mid-sixties, bad weather and the war with India again resulted in scarce supplies. The system of "provisioning" was instituted, whereby a subsidised, rationed amount of wheat was sold through the shops, with more available on the free market at a higher price. That system continued through the 1970s but was discontinued in 1987. [7]

The Zakat Programme

This is a unique and interesting experiment undertaken in Pakistan. It was initiated in 1980 under the Zakat and Ushr Ordinance by the government of Zia-ul-Haq, a right wing, conservative administration that probably saw the programme as an opportunity to shore up its Islamic credentials. At the time, there appeared to be no economic or special social protection motivations behind the programme's initiation. It is based on the Islamic injunction of charity, which mandates that all Muslims give 2.5 per cent of their wealth and assets each year to the poor. For many years, the government imposed a 2.5 per cent tax levy (a wealth tax) on specified financial assets of individuals such as bank deposits worth more than a certain amount. Zakat used to be deducted at source and siphoned to a Central Zakat Fund kept by the State Bank of Pakistan. However, since 1999 the Supreme Court has made the payment of zakat to the state voluntary.

Unlike the BISP, the Zakat Programme has no specific class of beneficiaries that it targets. The programme operates under the principle that deserving needy (mustahqieen) Muslims are to be given assistance. This particularly includes widows, orphans, the disabled and the unemployed (but not beggars). Two important issues about eligibility arise here: the programme, by design, discriminates on the basis of religion; and it has no a priori commitment to targeting by gender.

There are three types of programme, including Regular Zakat Programmes, Other Zakat Programmes and Special Zakat Programmes. The Guzara (Subsistence) Allowance is the principal programme and about 43 per cent of total funds were allocated to this in 2006/07. Beneficiary households received a monthly transfer of Rs. 500 under this scheme.

Regular Zakat Programmes also include education stipends; stipends for students of deeni madaris; health care; social welfare rehabilitation; and marriage assistance to women who are about to get married. Other Zakat Programmes include the Eid Grant and the Leprosy Patient Programme. Special Zakat Programmes consist of technical educational stipends.

Permanent rehabilitation grants, which started in 2001/02 with the objective of transferring larger lump sum grants of Rs. 10,000-50,000 to help the vulnerable address their problems on a permanent basis, have now been discontinued owing to lack of funds.

With payments to the programme now voluntary, and with large public mistrust, the programme's funding is in jeopardy, with both zakat levy collections decreasing and programme reserves drying up. Contributions to the Zakat Programme have observed a substantial decrease, from 0.3 per cent of GDP in the 1980s to 0.08 per cent in 2002-03.

Despite being one of the two largest safety net programmes in Pakistan along with the Food Support Scheme of the Pakistan Bait-ul-Mal, only 3 per cent of all households receive benefits, mostly under the Guzara Allowance (2.2 per cent) and permanent rehabilitation grants (0.4 per cent).

In mid-2009, a reform effort was initiated, sponsored by donors such as the Japan International Cooperation Agency (JICA) and driven in part by the focus on social protection by way of the BISP. The Minister for Religious Affairs, Zakat and Ushr acknowledged that the programme suffered from 'wrong perceptions of misappropriation and management in the system. To better target the programme to the poor, needy and destitute, the following reforms were initiated in September of the same year:

- Introduction of a new computerised mechanism to monitor and oversee zakat collection and disbursements;
- Central Zakat Council to be brought under the chairmanship of a serving judge of an apex court who would play an authoritative role in the distribution of donations;
- Ministry to prepare a revised budget for the programme under which major reallocations were made away from the Guzara Allowance to other sub-programmes: 50 per cent to education; 25 per cent to health; and 15 per cent to social welfare;
- Guzara Allowance to be discontinued
- Microfinance initiatives to be undertaken in collaboration with the State Bank of Pakistan;
- Hospitals to receive a 'huge amount' from the ministry to treat deserving patients;
- Ministry to give life insurance to one member (head) of beneficiary families;

The above-mentioned reforms are very much in the initial stages and it remains to be determined how many of the proposed changes will actually be implemented on the ground. [10]

Tackling Child Malnutrition

The National Nutrition Survey 2011 found that 15.1 per cent of children under five were wasted and 43 per cent stunted. It concluded that there had been no progress on improving the state of nutrition in the past decade.

A Nutrition Wing was then established in 2011 within the Ministry of Health and a Nutrition Section was set up in the Planning Commission. However, nutrition still lacks support or focus, exemplified by the dissolution of the National Nutrition Programme, which was passed by the Planning Commission but failed to get final government approval.

Several interventions related to nutrition have been implemented such as vitamin A supplementation, wheat flour fortification and salt iodisation, but these have largely been stand-alone programmes that have been implemented in isolation rather than as a complementary nutrition package. Support for mother and child health has been partially addressed through the Lady Health Workers scheme introduced by the government to promote good health and provide basic services to women at village level.

The overall humanitarian nutrition response has focused primarily on short-term emergency treatment programmes for acute malnutrition with little focus on addressing the underlying causal factors and on establishing long-term, sustainable and preventive programmes through the provincial and national governments. [6]

National Zero Hunger Action Plan (NZHAP) of Government of Pakistan

The NZHAP was inaugurated by Pakistan government in 2012 to address the problems emerging from high prices of food commodities and its impact of poor segments of the population. The NZHAP, which was developed by stakeholders' consultation, aims at reaching 61 million people in Pakistan, who are affected by hunger and malnutrition with a financial package of $ 16 billion, within five years. The NZHP comply the "right to food" for every citizen of Pakistan.

World Food Programme and Ministry of National Food Security have also signed an agreement, under which some 500,000 metric tonnes of wheat, donated by government of Pakistan will be converted by WFP, into nutritious food commodities that will be used to combat hunger among the most vulnerable sections of society particularly malnourished children, pregnant women and primary school children. The programme also features school feed programme and establishment of "zero hunger shops" in 45 extremely food insecure districts in Pakistan.

Some of the other salient interventions of Zero Hunger Plan include, (1) Support to food insecure households, (2) Cash/food support in disaster hit areas, (3) expansion of farm outputs and market access, (4) Targeted social safety nets, (5) Rationalisation of market prices of food commodities, (6) Improved nutritious quality of food intake (fortified food), (7) Diversification of food, (8) Food processing industry even at community level, (9) Food and nutrition awareness and education programme, (10) food security surveillance, (11) research for food policy/production, quality improvement, storage, distribution, and (12) Enhanced coordination among various federal and provincial ministries, and public-private-civil society partnerships.

The newly created ministry of national food security has not been able to implement its 'Zero Hunger' project even after seven months of its launch by former Prime Minister Yousuf Raza Gilani. No formal approval has yet been accorded to the budgetary requirements of the zero hunger project. With Gilani's exit, the project's pace has slowed down either due to the new ministry's lack of interest or lack of cooperation from other ministries. [2], [3]

Special Programme for Food Security (Crop Maximisation Project-I -CMP-I)

Keeping in view the importance of agriculture and ensuring food security on sustainable basis, the government of Pakistan started a Special Programme for Food Security (SPFS) with major support from FAO.

The SPFS project was piloted initially in three villages including two sites in Punjab and one in Khyber Pakhtunkhwa with twin objectives:

(1) To ensure the adequacy and access to food supply.

(2) To maintain the stability in agricultural production.

The major focus of this programme was on enhancing agricultural productivity and profitability of the major food crops-wheat, rice, maize and oilseeds, on a sustainable basis. The project was first up scaled to 109 villages in May 2002 under the name of Crop Maximisation Programme (CMP-I) and then extended to 1012 villages all over Pakistan in 2008 (called CMP-II), totally sponsored by the Government of Pakistan. The results of the SPFS were encouraging at two sites in Punjab on the basis of which 109 villages brought under this net in the name of CMP-I. The performance of CMP-I was dismal. Despite that the CMP-II was initiated in 1012 villages in Pakistan with a target to extend it to 13000 villages. [14]

The Special Programme for Food Security and Productivity Enhancement for Small Farmers (Crop Maximisation Project-II - CMP-II)

The main objective of this ambitious project is to supplement country's ongoing efforts to increasing food production through enhancing crop and livestock productivity. The CMP-II is a Rs. eight billion five-year project. It aims at enhancing food production through supporting small farmers. The project covers 1,012 target villages in 28 districts of Pakistan. The government has partnered with development organisations for the implementation. Rural Support Programme Network (RSPN), which is a network of rural development organisations, is the official partner. The project involves social mobilisation and capacity building of farmers' organisations.

The project was approved by the government of Pakistan on May 4th 2007 for a period of five years (July 2007 to June 2012) with a total cost of Rs 8.013 billion. The main idea was to increase production of small farmers by 30-35 per cent through capacity building and technology packages.

The project includes organising small farmers into village organisations and facilitates them with revolving funds at village level for providing production loans for farm inputs and micro-credits. The project also aims to link the small farmers with the buyers through Input Sales marketing Centres (ISMCs). Each ISMC is planned to serve a cluster of 15-20 villages; it would help them in getting quality farm inputs and marketing their produce.

The project covers total 50,000 farmers in 1,012 Villages including 430 in Punjab, 230 in Sindh, 160 in KP, 150 in Balochistan, 40 in AJK and two in FATA. It serves holders of 5-25 acres of land. In addition to the ones mentioned above, the project provides following services;

- Establishment of 1012 Revolving Fund Accounts of Rs two to four million per village, for VOs to get credits for their member farmers.
- Establishment of livestock centres 130-150 for a cluster of five-eight villages
- three-four small dairy farms per village
- Establishing 10,000 small enterprises, such as fish farms, bee farms, sheep rearing
- Development of computerised farmers' data base

The future of the project is in the doldrums after the devolution under the 18th constitutional amendment, as government has refused to provide funding through federal set up, and projects are handed over to the respective provincial governments. At present about Rs. 500 million is required by the project for 2012- 2013. [9], [11]

Benazir Income Support Programme

- In the year 2007–08, the sharp rise in oil prices and primary products in the international as well as domestic market resulted in double digit inflation, which almost halved the purchasing power of the people. Hence, there was an urgent need for direct and speedy relief to the underprivileged sections of the society. Benazir Income Support Programme (BISP) is the Government of Pakistan's response to the said compulsions.

- BISP was initiated in October 2008 with an initial allocation of Rs.34 billion (US $ 425 million approximately) for the financial year 2008-09.
- The Programme aimed at covering 3.5 million families in the financial year 2008-2009.
- The allocation for the financial year 2012-13 is Rs. 70 billion to provide cash assistance to 5.5 million families, which constitutes almost 18 per cent of the entire population. Thus the Programme aims at covering almost 40 per cent of the population below the poverty line.
- The enrolled families are paid cash assistance @ Rs.1200 per month on quarterly basis; apart from benefits provided under graduation strategy like: long term interest free loans, vocational and technical training, health and life insurance coverage.
- BISP is being implemented all across Pakistan including: all four provinces (Punjab, Sindh, Baluchistan, and Khyber-Pakhtoonkhwa), Federally Administered Tribal Areas (FATA), Azad Jammu and Kashmir (AJK), Gilgit Baltistan (GB) and Islamabad Capital Territory (ICT).

A unique feature of the BISP is that it is by far the most gender-sensitive safety net intervention undertaken in Pakistan. The prime beneficiaries are women. However, for this, the delivery mechanisms of cash transfer need to change: Both the smart card and the phone banking are likely to make it difficult for women to participate fully in the programme. BISP Secretariat does not work with or involve the Ministry for Women's Affairs in design or implementation. BISP does not also initiate consultative decision making with women participants, NGOs working on women's issues, gender activists, etc. The programme also appears to have no coordination efforts underway with agencies administering other safety net programmes, such as the Zakat Programme, or the multi-billion Food Support Scheme in Punjab.

The actual implementation details are not available with us; nor do we have any evaluation studies on this scheme. Effectiveness of cash transfers, as discussed in the chapter on Sri Lanka, is dependent on many factors. We do not know how these factors play out in the provinces of Pakistan. [4], [5]

Bait-Ul-Mal

Pakistan Bait-ul-Mal is a semi-autonomous organisation within the Ministry of Women Development, Welfare and Special Education with a mandate to provide social assistance to the deserving poor and needy people including the elderly.

Bait-ul-Mal benefits are open to all regardless of creed and the funds are entirely controlled by public servants. It has offices at provincial and district levels that are closely linked to, but not part of provincial administrations. The Pakistan Bait-ul-Mal is headed by a Managing Director who reports to Board of Management about its progress. At provincial and district levels, there are Steering Committees to manage funds.

The scheme consists of the following programmes:

(a) Food Support Programme

The largest programme under Bait-ul-Mal is the Food Support Programme for the poor that provides Rs. 2400 a year in two six-monthly instalments. The programme guidelines require that assistance be provided to "needy individuals having no support or source of income" in the following order of priority:

(a) Individual with major ailments/disability;

(b) Widow with dependent children;

(c) Invalid with dependent children;

(d) Infirm (senior citizens above);

(e) Poorest of the poor to be reviewed periodically for rehabilitation; and

(f) Orphans, destitute and victims of unpredictable circumstances.

(b) NGO Funding

Bait-ul-Mal also provides Rs 43 million for 78 NGO projects to support institutions for disabled, orphans, abandoned and destitute women and other poor needing care as well as assistance for health (eye care) of aged persons, particularly from rural areas. The funding is based on cost-sharing-Bait-ul-Mal provides 75 per cent in rural areas and 50 per cent in urban areas. Approximately, 70 per cent of the expenditure goes to health facilities. Overall, Pakistan Bait-ul-Mal is estimated to have covered almost two million households by its activities. There is an overlap between Bait-ul-Mal and Zakat in the area of assistance to individuals. There seems to be no good reason for continuing the two agencies providing the same type of assistance to the same target group.

It would seem better to divide the responsibilities more clearly so that Bait-ul-Mal would deal with minorities or specific target groups only and this option needs to be seriously considered. The rules of both Zakat and Bait-ul-Mal require that a person assisted by one cannot get assistance from the other.

Bait-ul-Mal is currently developing a database, which hopefully will facilitate crosschecking of recipients. The more difficult issues relate to clerical procedures, such as the need for standard formats for names and addresses; deciding which system has priority if matching reveals different data; and ensure the accurate and timely information on the number and types of beneficiaries. [15]

European Union Food Facility

In May 2009, FAO launched a 26-month project to help the government of Pakistan restore the food security of vulnerable households in the food-deficient areas of North West Frontier Province, Punjab, Sindh and Balochistan. With funds from the European Union totalling € 24.7 million, the project will enable at least 166,700 farming families (reaching over one million people) to grow enough staple foods–wheat, rice, vegetables and other crops–to meet household food needs. FAO will provide farmers with free quality inputs, including high-yielding and locally appropriate seed varieties and fertilisers, in time for the planting season to help boost crop production.

Irrigation is crucial to agricultural activities in the targeted areas, yet several irrigation schemes have fallen into disrepair due to neglect and natural disasters. FAO will help rehabilitate 150 small-scale community-managed irrigation schemes and 200 rainwater harvesting structures.

FAO will also provide 100 seed and grain storage facilities to help reduce losses, as well as agricultural tools and machinery to 100 farmers' associations. To enhance sustainability, FAO will help train around 37 500 smallholder farmers in the efficient use of irrigation, crop production, proper maintenance and use of agricultural tools and machinery and post-harvest management. Training will also be provided through Farmer Field Schools.

The project, in line with the government's ongoing national programme for food security, will be implemented jointly by FAO and the UN World Food Programme (WFP). WFP will provide food-for-work assistance to approximately 83 000 households living in the most food vulnerable areas. [16]

Conclusion

Except for procurement and disbursement of wheat scheme, which is reported to have been ineffective because of lack of consistency, Pakistan does not seem to have tried any scheme for food security on a long term basis. It has some cash transfer programmes, like Bait-ul-mal, but we do not know the impacts generated by them. Programmes to improve agriculture technology have been implemented only intermittently. Consequently, self - sufficiency in food is not guaranteed. India has tried out many more programmes, and the effect can be seen in better performance of its indicators, in spite of having a much larger population to deal with.

References

The references are ordered alphabetically in the following list but the numbers given are according to the order in which the references appear in the text.

4. Benazir Income Support Programme official website - http://www.bisp.gov.pk/

11. Crop Maximisation Support Programme. http://www.ifad.org/operations/pipeline/pi/pak.htm

16. FAO - European Union Food Facility http://www.fao.org/europeanunion/eu-in-action/eu-countries/pakistan/en/

7. Household Food Security in Pakistan : The Ration Shop System - Pakistan Institute of Development Economics, International Food Policy Research Institute.

19. International Food Policy Research Institute.
http://www.slideshare.net/m-hameed/4-3-paul-dorosh-wheat-procurement-policy-dec-13-2012-final-updated

15. Pension and Social Security Schemes in Pakistan : Some policy options - PIDE

2. No progress on Zero Hunger Project, InPaper Magazine.
http://www.dawn.com/news/770158/no-progress-on-zero-hunger-project

3. Sustainable Development Policy Institute official website.
https://www.sdpi.org/policy_outreach/event_details164-2012.html

8.9. Scoping Study on Food Security status in Pakistan

5.10. The Benazir Income Support Programme and the Zakat Programme - A Political Economy Analysis of Gender, Shanza N. Khan and Sara Qutub

6. Tackling the Child Malnutrition in Pakistan -
http://everyone.savethechildren.net/

12. United Nations Statement on Food Security in Pakistan - The United Nations Systems in Pakistan.

13.14. The State of Food Security in Pakistan:Future Challenges and Coping Strategies, Pakistan Development Review.

1. Wheat Markets and Price Stabilisation in Pakistan : An analysis of Policy Options, Paul Dorush and Abdul Salam.
http://www.pide.org.pk/pdf/Working%20Paper/PIDE % 20 Working % 20 Papers % 202006-5.pdf

17. Wheat Policy - Problems and Issues, Mohhamad Khalique Arshad.
http://www.pakissan.com/english/issues/wheat. policy. problems. and.issues.shtml

18. Wheat subsidy - a blessing or curse?, Adnan Falak.
http://www.nation.com.pk/columns/12-Jul-2012/wheat-subsidy-a-blessing-or-a-curse

11

Mongolia

Compilation: Nikhil Pritmani

Introduction

Mongolia is a landlocked country in east-central Asia. Bordered by Russia and China, Mongolia is a parliamentary republic. Its population is 2.796 million (2012) and GDP is US$10.27 billion (2012).

Mongolia has been famous for the great ruler Genghis Khan who tried to rule the world. But this is not the only history of Mongolia. Medieval Mongolia had long been under rule of Chinauntil early 20th century when Outer Mongolia declared independence from China on November 30, 1911 under the leadership of Bogd Khan. In 1919, after the October Revolution in Russia, Chinese troops led by XuShuzheng occupied Mongolia. After several conflicts with China and with the help of Russia, Mongolia again declared independence on July 11, 1921. Thus Mongolian People's Republic was created with Mongolian People's Party as the ruling government. Bogd Khan was the nominal Head of the state.

After death of Bogd Khan, who opposed the Soviet assistance, in 1924 the country came under Soviet influence and Stalin's orders of collectivisation of agriculture as well as liquidation of Buddhist institutions were forcefully carried out. Mongolian People's Republic became a Soviet satellite state in spite of much resistance. About 30,000 people lost their lives in this struggle.

In the summer of 1945, the Soviet Union used Mongolia as one base for launching a successful attack against the Japanese. The preceding build-up brought 650,000 Soviet soldiers to Mongolia, along with massive amounts of equipment. The Mongolian People's Army played a limited support role in the conflict, but its involvement gave Stalin the means to force the Chinese side finally to accept Mongolia's independence. Outer Mongolia, for all practical purposes was not under China, but on paper, it was still considered a part of China. In 1945, China reluctantly gave in to the demands of making Outer Mongolia independent although the Inner Mongolia remained with China. In spite of close integration with Soviet Russia, Mongolia was an independent nation and was made the member of UN in 1961.

After Gorbachev came to power, on July 29, 1990, multi-party elections were held in Mongolia. Mongolia People's Republic Party came to power. Democracy was introduced with a new constitution [9]. After independence of Mongolia and introduction of capitalism, the economy and agriculture of Mongolia has seen a decline. Also several dzuds affected the agriculture deeply and thus not having any other reliable source, agriculture and husbandry along with economy had seen a decline. However these started to revive with the incoming of 21st century and development of mining and service sector of Mongolia.

Mongolia is a lower middle income economy as declared by World Bank. 22.4 per cent of its population lives below US$ 1.25 per day (Human Development Report 2011). Life expectancy at birth is 67 years (2012) [10]

About Mongolia:

Area:
total: 1,564,116 km^2
land: 1,553,556 million km^2
water: 10,560 km^2

Coastline: 0 km (landlocked)

Land use:
arable land: 0.39%
permanent crops: 0%
other: 99.61% (2011)

Irrigated land: 843 km^2 (2011)

Total Renewable Water Resources: 34.8 km^3 (2011)

(Ref: *http://en.wikipedia.org/wiki/Geography_of_Mongolia)*

The terrain is one of mountains and rolling plateaus, with a high degree of relief. Mongolia is high, cold, and dry. It has an extreme continental climate with long, cold winters and short summers, during which most precipitation falls.

Table 1 gives a few details that indicate current position of Mongolia vis-à-vis several other similar countries in the world.

Table 1: Comparison of Development Indicators of Some Asian Nations

Country	HDI	Multi-Dimensional Poverty Index	Global Hunger Index	GDP nominal per capita ($)	Global Food Security Index
India	0.554	0.283 (2012)	21.3	1504	45.0
Bangladesh	0.515	0.292	19.4	903	35.9
Sri Lanka	0.715	0.021	15.6	3161	49.5
China	0.699	0.056	5.5	6747	61.4
Pakistan	0.515	0.264	19.3	1307	40.2
Burma	0.498	0.154	N/A	868	46.6
Vietnam	0.617	0.084	7.7	1901	49.4
Cambodia	0.543	0.251	16.8	1015	31.9
Mongolia	0.675	0.065 (2012)	10.8	3971	N/A
Iran	0.742	N/A	<5	4750	N/A
Jordan	0.700	0.008	<5	5174	52.3
Thailand	0.690	0.006	5.8	5674	59.2
Indonesia	0.629	0.095	10.1	3509	45.9
Malaysia	0.769	N/A	5.5	10547	63.6
Lao PDR	0.543	0.267	18.7	1476	N/A
Nepal	0.463	0.350	17.3	692	34.4

(Ref:*http://www.ifpri.org/sites/default/files/publications/ghi13.pdf*

http://hdr.undp.org/sites/default/files/hdr2013_en_summary.pdf

http://en.wikipedia.org/wiki/List_of_countries_by_GDP_%28nominal%29_per_capita

https://data.undp.org/dataset/Human-Development-Index-HDI-value/8ruz-shxu

https://data.undp.org/dataset/MPI-Multidimensional-poverty-index-/746y-qwam)

As can be seen from the figures above, Mongolia is ahead of India in all the food security indicators for which data is available. Its per capita income is almost three times as much as India and the hunger levels are almost half. It is more than hundred points ahead of India in HDI. However, the situation might have worsened for Mongolia over the last decade. We investigate that in our chapter.

Demographic Indicators

India has almost 450 times more population compared to Mongolia (table 2). And its growth rate of population has been decreasing at a slow rate and hovers around 1.3 to 1.5 per cent (table 3). For Mongolia, the population growth is steady at the rate of 1.5 per cent in most years (table 3). Thus the population growth rates are comparable.

In other demographic figures also, Mongolia seems to be at the same level as India. The life expectancy at birth in Mongolia is 67.98 while in India it is 66.46. The annual number of births per 1,000 people in Mongolia is 21.03 while in India it is 21.34.

Table 2: Population Comparison.

Year	India	Mongolia
2011	1,221,156,319	2,754,209
2001	1,059,500,888	2,419,669
1991	886,348,712	2,217,937

(Ref: *http://data.worldbank.org/indicator/SP.POP.TOTL)*

Table 3: Average Annual Population Growth Rate.

	1991	2001	2011
India	2.0	1.6	1.3
Mongolia	1.5	0.9	1.5

(Ref: *http://data.worldbank.org/indicator/SP. POP. GROW/countries/ MN?display = default)*

About 47.07 per cent (2013) of the population lives in Ulaanbaatar, the capital city of Mongolia. Mongolian economy is primarily based on Mining, livestock breeding and agriculture.

Arable land and permanent crops cover 1.3 million ha, while permanent pastures cover 117.1 million ha. The main crops grown in Mongolia are wheat, barley, pulses, potatoes and some fruits. The livestock sector dominates, contributing 84.9 per cent of total agricultural production.

Economic Indicators

Table 4 shows that GDP of Mongolia has been raising constantly except during 2009 when it was hit by dzud very hard. Both India have Mongolia are developing economies. Mongolia has achieved a high GDP growth rate due to recent development in industrial and mining sectors.

Table 4: GDP Comparison

Year	India	Mongolia
2013	1,758,000,000,000	11,140,000,000
2012	1,841,709,755,679	10,271,393,281
2011	1,872,840,195,945	8,761,426,371
2010	1,710,908,768,960	6,200,357,070
2009	1,365,372,433,272	4,583,834,427
2008	1,224,095,295,077	5,623,236,708
2007	1,238,700,195,725	4,234,894,168
2006	949,116,769,688	3,414,053,251
2005	834,215,013,543	2,523,359,941
2004	721,585,608,172	1,992,066,759
2003	618,356,467,439	1,595,297,301
2002	523,968,561,872	1,396,555,772
2001	493,954,333,969	1,267,997,923
2000	476,609,148,198	1,136,896,162
1999	466,867,079,633	1,057,408,608

(Ref: *http://www.factfish.com)*

The semi-log regression, with log of GDP as the dependent variable and time as an independent variable shows that the Indian nominal GDP grew at the rate of 11.65 per cent between 1999-2013 (R square = 0.967 and t statistics was 19.54). The nominal GDP grew at 18.24 per cent for Mongolia (R square = 0.98 and t statistics = 25.4).

However, the nominal GDP is very misleading. As shown in table 5, the average rate of inflation was 9.38 per cent for Mongolia (with standard deviation of 5.53) whereas India had the average inflation rate of 6.68 per cent for the same period (with standard deviation of 2.89). Thus the growth rate of Mongolia would be still higher to India's but not as high as the nominal growth rate shows. Also, due to intermittent periods of very high inflation, Mongolia's economy seems to be more unstable compared to India. On the one hand, the economy has taken off well with high growth but on the other hand, inflation seems to be coming in the way of it translating into well-being of the citizens. There are often complains that prices of food products double over a short periods which seems a big reason behind high poverty figures of this developing nation.

Table 5: Inflation Comparison

Year	India	Mongolia
2013	9.6	8.2
2012	9.3	15.0
2011	8.9	9.5
2010	12.0	10.1
2009	10.9	6.3
2008	8.4	25.1
2007	6.4	9.0
2006	6.1	5.1
2005	4.2	12.7
2004	3.8	8.2
2003	3.8	5.1
2002	4.4	0.9
2001	3.7	6.3
2000	4.0	11.6

1999	4.7	7.6
1998	13.2	9.4
1997	7.2	36.6
1996	9.0	46.9
1995	10.2	0.0
1994	10.2	87.6
1993	6.4	268.2

(Ref: *http://www.factfish.com)*

Poverty Ratios

Interestingly enough, Mongolia has higher proportion of population living below the poverty line, compared to India. As explained earlier, this might have happened because Mongolia is still recovering from the effects of its separation from Russia and its unstable economy resulting in volatile inflation rates. Mongolia had much lower poverty figures than India in 2005 (as given in Table 1), but the situation seems to have worsened over years.

Table 6: Poverty Headcount Ratio at National Poverty line (Per cent of Population)

In Mongolia	
2012	27.4%
2011	33.7%
2010	38.7%

(Ref: *http://data.worldbank.org/country/mongolia)*

In India	
2009-10	29.80%
2011-12	21.92%

(Ref:*http://www.rbi.org.in scriptsPublicationsView.aspx?id=15283)*

Health Indicators

As Table 7 shows, the IMR in India declined by 23 per cent between 1995 and 2010. The decline was faster for Mongolia: the IMR declined by about 34 per cent. In 2010, the IMR of Mongolia (36.03 per cent) was much lower than that of India (52.91 per cent).

Table 7: Infant Mortality Rate: Death of Infants Per Thousand Live Births.

	1995-2000	2000-2005	2005-2010
India	68.86	60.70	52.91
Mongolia	54.70	43.54	36.03

(Ref: *http://data.worldbank.org/indicator/SP.DYN.IMRT.IN)*

Table 8: Maternal Mortality Rate.

	1995	2005	2013
India	460	280	190
Mongolia	120	89	68

Ref: *http://data.worldbank.org/indicator/SH.STA.MMRT*

Table 8 shows that the MMR of Mongolia (68) is just one third of that of India (190). The explanation might lie in two factors:

1. Per capita health expenditure is much more in Mongolia even in recent years. The economy might be faltering on some counts, but the health sector seems to be still given importance. As said earlier in the first chapter, investment in social sectors goes a long way in improving the living standards of a population.
2. 97.4 per cent of Mongolia's population is literate. In India, this proportion is 72 per cent. It has been a well-established fact that educated people have more "desirable" reproductive behaviour compared to the ones who are not educated. (Ref: http://www.indexmundi.com/mongolia)

One more reason can be the health expenditure per capita made in the two nations. Mongolia spends almost four times or 400 per cent more on health compared to India (table 9). No wonder that the health indicators are better!

Table 9: Health Expenditure Per Capita.

	2009	2010	2011	2012
Mongolia	111	143	190	232
India	44	52	62	61

Ref: *http://data.worldbank.org/indicator/SH.XPD.PCAP*

Malnutrition Prevalence (Per cent)

In 1994, the malnutrition picked to a level as high as 52.6 per cent of the population in Mongolia (table 10). That time India had 26.9 per cent malnourished. In last 20 years, India's figure has only marginally improved to 17.5 per cent being malnourished. But Mongolia halved its malnutrition percentage in the same period.

Mongolia had malnutrition levels that were almost the double of India's in some years between 1991 and 2011. For both the nations, the levels are declining over years, although the fall has been much faster for Mongolia compared to India. Still, in 2011, the level of malnutrition was significantly higher in Mongolia.

Table 10: Per cent of Population Malnourished

Date	India	Mongolia
2011	17.5	24.2
2010	18.3	25.6
2009	18.8	26.3
2008	19.0	27.6
2007	19.4	29.2
2006	20.2	31.2
2005	20.9	32.5
2004	21.4	33.4
2003	21.4	34.1
2002	21.8	34.8
2001	21.6	35.9
2000	21.3	37.6
1999	20.5	39.8
1998	20.8	41.9
1997	21.5	43.6
1996	23.1	45.6
1995	25.2	48.5
1994	26.8	52.6
1993	27.2	52.4
1992	26.3	47.5
1991	26.9	37.5

(Ref: *http://www.factfish.com/statistic-comparison/mn-in/malnutrition % 20 prevalence)*

Expenditure for Food Security

Mongolia: The National Food Security Programme embraces commercial as well as household food security and is budgeted at US$ 1,289.23 million over two phases: (i) 2009-2012 and (ii) 2013-

2016. The budget includes USD 500 million for the supporting Agricultural Investment Fund [4].

India: National Food Security Act, 2013 has a total expenditure of 1.122 trillion rupees (or between $ 20 and $ 21 billion). However, the population of India is also almost 450 per cent more than Mongolia's, so some change in size of budget can be expected. Still the figure is much higher and rationalisation of such expenditure can save huge sums of money. This is especially relevant because the malnutrition figures have been declining at much faster rate in Mongolia compared to India [8].

Some other Figures of Mongolia
GDP by Composition:

Agriculture: 14%

Industry: 29.9%

Services: 56.1% (2012 est.)

(Ref: http://www.indexmundi.com/mongolia)

Exports: USD$ 4.2727 billion (2013)

(Ref: http://www.wto.org/english/res_e/statis_estatis_bis_e.htm? solution=WTO&path=/Dashboards/MAPS & file=Map.wcdf & book mark State={%22impl%22:%22client%22,%22params%22: {%22 lang Param %22: %22en %22}})

Imports: USD$ 6.3547 billion (2013)

(Ref: *http://www.wto.org/english/res_e/statis_estatis_bis_e.htm? solution=WTO & path=/Dashboards/MAP S& file=Map.wcdf & book mark State ={% 22impl %22: % 22 client %22, % 22 params %22: {% 22lang Param % 22: % 22en % 22}})*

Food Security of Mongolia

History of Food Security

In 2005, approximately 11,600 km 2of the land in Mongolia (0.74 per cent) was classified as being arable or having permanent crops, a fairly small amount compared to the 1,293,000 km2 (83.5 per cent) classified as pasture land (FAO 2008). Approximately 1,780 km2of land was actually harvested for crops (15 per cent of the arable land, and 0.11 per cent of the total land area), the vast majority of which was harvested for wheat (86 per cent).

These numbers have varied significantly in the past. Although broad temporal patterns generally match those for production, two differences are notable. First, the 1980s were marked by large increases in production without dramatic increases in the amount of land harvested for crops, underscoring that increasing efficiency (e.g., better farm technology, more fertilisers) primarily drove the production increases seen during this time. Second, the collapse of production since 1990 has been accompanied by dramatic decreases in both the amount of land harvested and efficiency [1].

Soviet assistance, at its height one-third of GDP, disappeared almost overnight in 1990 and 1991 at the time of the dismantlement of the USSR. Since 1990, Mongolia as a process of entering in free market economy has been disintegrating the state-owned crop production enterprises into small private farms. But most of them did not have either capital or technological requirements to function properly. Severe drought and dzuds since 1998 for a few years was one more reason for financial weakness of farmers. Due to weak agriculture the prices of basic food products have also been increasing. Prices of bread, carrot and beef have more than doubled in a year whereas prices of potatoes have tripled. Such high increases in essential commodities have made life really tough for the Mongolians.

Table 10: Price Change in Essential Commodities

	2007 (US cents)	**2008 (US cents)**
Bread	25	57
Rice/kg	51	68
Flour/kg	42	77
Milk	42	77
Tea	21	38
Beef/kg	170	330
Horse Meat	150	205
Mutton/kg	130	240
Potato/kg	20	70
Carrots	30	70

(Ref: *http://www.worldvision-*

institut.de/_downloads/allgemein/Food%20security%20in%20Mongolia.pdf)

Need of Food Security in Mongolia

The agricultural sector comprises more than 80 per cent of the rural economy and employs more than 60 per cent of the rural population. From these, around 50 per cent of Mongolians work as herders. Thus agriculture and animal husbandry are the key strategic sectors for rural development. In addition, the agricultural sector contributes 17.1 per cent to Mongolia's Gross Domestic Product, of which nearly 85 per cent is derived from livestock husbandry. Desertification, caused in part by over-grazing, is robbing inner Mongolia of farming land and pushing families to move towards cities in search of paid work. According to the government, 36 per cent of people are living on less than 68 US cents per day.

Extreme weather conditions make both crops and herds a volatile investment. In 2005 the worst wheat crop on record forced Mongolia to increase expensive imports further to ensure enough food was available. As the price of home-raised meat has increased beyond the reach of families, consumption of vegetables like potatoes and carrots has meant a greater demand for imported vegetables, and therefore prices rise in these ingredients also.

Some 70 per cent of Mongolia's food is imported from China and the rising Chinese currency and costs of transporting food are making it much more expensive for parents to feed their children. Children are already facing challenges of malnutrition in Mongolia. Around 13 per cent of them are underweight and many suffer from other nutrition-based conditions such as skin disease and rickets. The impact of rising food costs will be seen not only in increased malnutrition of children and their parents, but also in decisions made at a family level which affect children, for instance taking them out of school or sending them into exploitative labour [3].

Policy Decisions for Food Security

Mongolia aims at increasing incomes and creating better livelihoods for vulnerable rural populations. For this they need:

(1) Improved productivity of herders and farmers

(2) Improved access to markets for herders and farmers

(3) Improved livelihood security for herders and farmers.

In relation to increasing the productivity of farmers, emphasis will be placed on improving animal health, providing vegetable seed, and the establishment of extension systems and market linkages.

Building upon its past achievements, Mongolia will continue to invest in the improvement of veterinary policy and associated legislation, as well as in the control of animal diseases and the strengthening of veterinary education through its Animal Health Project. At present, brucellosis and foot-and-mouth disease are hindering the development of Mongolia's meat exports.

Food Security Programmes of Mongolia

Special Programme for Food Security (SPFS)

The first security programme, from 2001-2007, was an ambitious endeavour to address weaknesses in the food supply chain to provide Mongolians with sufficient, safe and nutritious food and to mitigate the risk posed by over-reliance on imported food staples. This is because Mongolia is experiencing a price escalation of food staples and serious food safety incidents.

The Special Programme for Food Security (SPFS) in Mongolia initially consisted of two parts: (i) small-scale irrigation rehabilitation and horticulture demonstration, and (ii) cereal sub-sector demonstrations. The first part became operational in November 1997.Horticultural activities were promoted on four sites in the North-Central region. Meanwhile, the improved cereal production technology project was approved with the objectives (i) to train farmers in multiplying quality seed on the basis of farmer saved seed; (ii) demonstrate improved fallow technologies and other conservation farming practices; (iii) train farm managers and farmers in using conservation farming practices to improve cereal production.

The project ended in May 2005 having made significant in-roads in promoting reduced tillage systems, although it has been affected by adverse climatic conditions (harsh winters of 2000 and 2001, and a dry summer in 2000). In 2004, the SPFS began focusing on increasing the supply of dairy foods by reducing post-harvest losses and restocking.

National Food Security Programme (NFSP)

The Government of Mongolia approved the National Food Security Program (NFSP) for 2009-2016. The overall goal of the programme is to provide the entire nation with stable supplies of accessible, nutritious and safe food to create healthy livelihoods and high labour productivity. The programme emphasises the participation of the people, the government, and the public and private sectors.

The Programme focuses on:

- Food Production and Supply
- Food safety
- Nutrition
- Legal environment, organisation and funding

The Programme is implemented in two phases:

- First Phase: 2009-2012
- Second Phase: 2013-2016

The following are the objectives of the programme:

- Create enhanced enabling legal, economic and organisational environment for ensuring food supply, quality and safety:
- Revise the laws and regulations related to food security, quality and safety to comply with short- and long-term development objectives;
- Create enabling economic, financial and business environments to increase foreign and domestic investment;
- Build adequate management structure and inter-sectoral coordination and human capacity to implement the programme; stable supply of the population with nutritious, secure and accessible foods and increase the proportion of the industrially processed food in overall consumption:
- Enhance supply of staple food by domestic production to substitute imports;

- Reduce disparities in food access between urban and rural areas and seasonality of supply;
- Stabilise food supply of the population, especially to the people of the vulnerable groups, during natural calamities, economic instability and food deficits;
- Update food processing technologies to enhance market competitiveness of the industry and export capacity of the Mongolian brands of food products;
- Improve monitoring and information network to secure hygiene and safety of food
- Products and drinking waters:
- Ensure quality and safety of food products through introducing GAP and GHP at all stages of food preparation, processing, preservation, transporting and marketing;
- Improve monitoring and information network of food quality and safety, and related practices;
- Improve access, quality and safety of drinking waters;
- Improve nutritious quality of food, supporting adequate, healthy diets and reduce nutrition deficiency, preventing from risk factors of non-communicable chronic diseases;
- Elaborate and implement updated recommended dietary intake norms of a Mongolian person, improve consumer education of the population on food nutrition;
- Monitor and prevent from micronutrient deficiency and non-communicable chronic diseases;
- Increase production and consumption of fortified and functional food products.

An Advisory Council in charge of implementation of the Programme throughout the country will be established to organise and monitor the implementation of the Programme.

The following financial resources will be used for implementation of the Programme:

- Own resources of business enterprises and residents;
- State budget;

- Local budgets;
- Banks' loans;
- Aids and soft loans of foreign countries and international organisations;
- Investments of domestic and foreign investors.

Budget Summary

Pillar/Priority Area 1: Enabling Environment

Title of sub-project	Total Cost, US$ million
1.1 Capacity Building: Vocational Training for the Food and Agriculture Sectors	13.1
1.2 Capacity building of food laboratories	14.0
1.3 Promotion of Food Research, Information and evaluation	8.5
1.4 Agricultural Investment Fund (newly suggested)	500,0
Sub Total	**535.6**

Pillar/Priority Area 2: Food Security

Title of Sub-project	Total Cost, US$ Million
2.1 Meat production	16.4
2.2 Milk production	40.1
2.3 Poultry production	1.96
2.4 Fish production	1.2
2.5 Crop diversification	15.0
2.6 Irrigated crop production	187.48
2.7 Renovation of crop equipment	178.82
2.8 Fertiliser	3.75
2.9 Plant protection	4.3
2.10 Vegetables and Potatoes	12.0
Sub Total	**461.01**

Pillar/Priority Area 3: Food Safety

Title of Sub-project	Total Cost, US$ Million
3.1 Introduction of modern inspection system into food production and processing	3.0
3.2 Integrated of agricultural and processing factories	190.0
3.3 Prevention of food produce from physical, chemical, biological and mechanic contamination and improvement of inspection	2.0
3.4 Food reserve	19.83
3.5 Prevention of food borne disease	1.50

3.6 Set Up "Mongolian Brand Name"	1.80
Sub Total	**218.13**

Pillar/Priority Area 4: Nutrition

Title of Sub-project	TotalCost,US$ Million
4.1. Improve the nutrition education of the population	12.6
4.2. Strengthening the capacity of nutrition scientific research work and Information system	6.5
4.3. Prevention of micronutrient deficiency	4.8
4.4. Increase of Production and Consumption Fortified Food with Minerals and Vitamins	4.9
4.5. Enhancing capacities for preventing from non-communicable diseases food related risk factors	1.2
4.6. Food Access Improvement	0.585
4.7 Drinking water supplies	43.9
Sub Total	**74.485**

Thus total Budget of NFSP is 1289.225 US$ million (including Agricultural Investment Fund).

(Ref: *http://gafspfund.org/sites/gafspfund.org/files/Documents/ Mongolia_5_of_ 9% 20 National%20 Food%20 Security%20 Programme. pdf)*

Conclusion

Mongolia was faring better than most South Asian and South East Asian nations till it was under the Russian control. After gaining independence from Russia, Mongolia actually has lost its economic momentum. Lack of knowledge about agriculture technology and droughts coupled with lack of timely credit has led to reduction in supply of agricultural products, leading to galloping inflation. Food imports have started becoming expensive. In spite of all these problems, riding on the mining sector, the Mongolian economy has been growing very fast. Its malnutrition figures also have been declining - we feel that it must be mainly due to the increased urban incomes. Thus economic growth seems to have brought in some notable results for Mongolia. The rural areas still struggle for survival and the malnutrition rates there seem to be high. Mongolia has decided to concentrate on agriculture and allied sectors development to solve the problem. It is not going for any direct feeding programmes.

There are many contrasts between India and Mongolia. Mongolia has been able to build its economy based on the firm social sector foundation. India did not create such a strong social sector base. The density of population is very low in the rural areas in Mongolia and the arable land is extremely limited. Mongolia's growth has to rely on mining and industrial growth more. With its large rural population, and a well- developed agriculture sector, India would like to keep a sizeable population in rural areas. But in spite of such differences, the most important thing to learn from Mongolia is to understand how in spite of such a strong shock to the economy, Mongolia has been able to keep its malnutrition rates under check. Once the population has been enabled, it is easier for the nation to face serious challenges also.

References

1. Didier and Lkhamjav (2009), "The Potential for Intensive Crop Production in the Eastern Steppe of Mongolia: Report prepared for WCS TransLinks Program, USAID http://s3. amazonaws.com/WCS Resources file_ 20110518 _073805 _Case Study _Intensive Crop Production Potential- Mongolia_ DWWW. pdf
2. Factfish World Statistics http://www.factfish.com/
3. FAO Corporate Document Repository, "Mongolia" http://www. fao. org/docrep /005 /y 2722e/ y2722e0 y. htm
4. Government of Mongolia: National Food Security Programmehttp:// gafspfund.org/sites/gafspfund.org/files/Documents Mongolia_8_of_9_Consultations_Brief_Agriculture_Plan_NFSP. pdf
5. Government of Mongolia: "National Programme for Food Security: 2009-2016" Ulaanbaatar 2009 http://gafspfund.org/sites/ gafspfund.org/ files/Documents/ Mongolia _5 _ of _ 9 % 20 National % 20 Food % 20 Security % 20 Programme. pdf

6. Indexmundi: Mongolia http://www.indexmundi.com/mongolia
7. Wikipedia, "Geography of Mongolia" http://en.wikipedia.org/wiki/Geography_of_Mongolia
8. Wikipedia, "National Food Security Act" http://en.wikipedia.org/wiki/ National_Food_Security_Act,_2013
9. Wikipedia, "Zud" http://en.wikipedia.org/wiki/Zud
10. World Bank, http://data.worldbank.org/country/mongolia
11. http://data.worldbank. org/indicator/
12. World Vision, "Mongolia: Facing food challenges" http://www. worldvision-institut. de/_downloads/ allgemein/Food % 20 security % 20 in % 20 Mongolia. pdf
13. Reserve bank of India, Table 162 http://www. rbi.org.in/scripts/ Publications View. aspxid = 15283.

Attaining Food Security